CW01521935

CONSERVATION 2000

The Ozone Layer

© Philip Neal 1993
First Published 1993
All rights reserved. No part of this publication may be reproduced in any form or by any means without permission from the Publisher.

Typeset by J&L Composition Ltd, Filey, North Yorkshire

and printed in Hong Kong

for the publishers
B.T. Batsford Ltd
4 Fitzhardinge Street
London W1H 0AH

A CIP catalogue record for this book is available from the British Library

Acknowledgements

All information about the Sun Smart campaign and the illustrations on pages 6/7 and 23, were supplied © of the Anti-Cancer Council of Victoria, Australia. The author acknowledges the help of Dr Robin Marks, the consultant dermatologist of the ACCV, Richard Shield of ELE and David Muir of Bristol City Council.

The Authors and Publishers would like to thank the following for kind permission to reproduce illustrations: The Associated Press Ltd for page 8; FOA for page 49; Katz Pictures for pages 14/15, 22/23 and 57; Magnum for pages 42/43; Military Picture Library for page 13; National Railway Museum for pages 10/11; David Pratt for pages 32A, 32B and 53; Science Photo Library for pages 18/19, 20, 26/27, 34/35, 38/39, 46/47, 50/51 and 54/55; Frank Spooner Pictures for pages 30/31, 58A and 58/59; Stills Pictures/Mark Edwards for page 60; Tony Stone for page 24; WATCH Ozone Project for page 29a and 29b; ELE for page 19b and 28. All other pictures were obtained from the author. The drawings on pages 9, 17, 21, 33, 37A, 37B, 40, 41 and 44 were by Ken Smith. The pictures were researched by David Pratt.

CONSERVATION 2000

The Ozone Layer

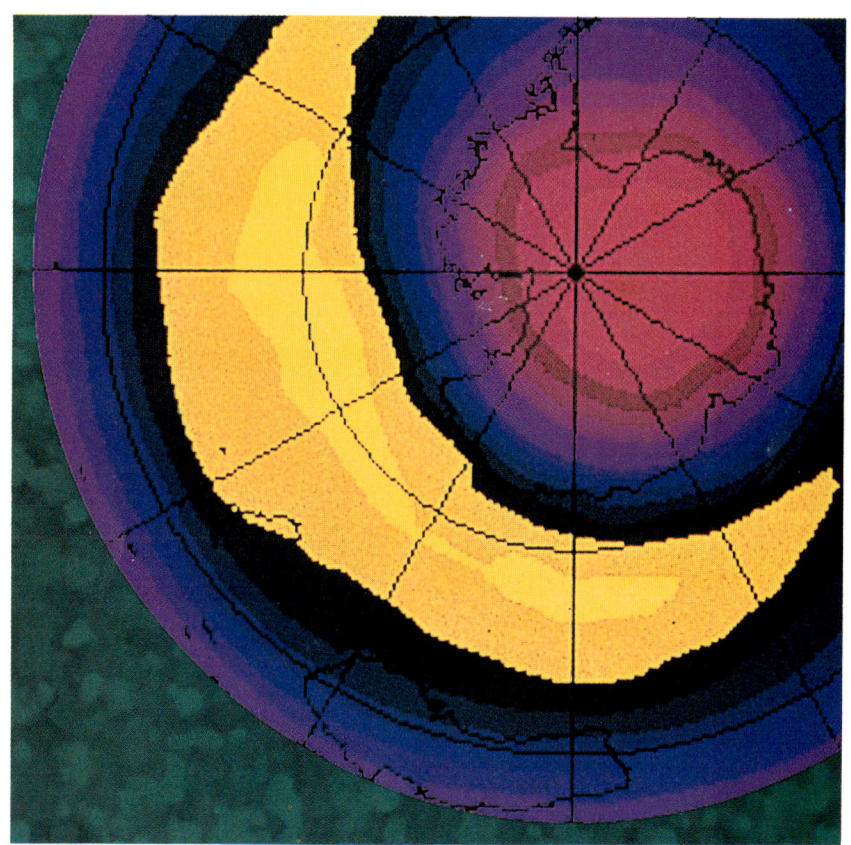

Philip Neal

B.T. Batsford Ltd London

CONTENTS

Gases — Chemical Symbols

For convenience, the symbols for gases are used throughout the book:

CFC	Chlorofluorocarbon	N_2O	Nitrous Oxide
CO	Carbon Monoxide	NOx	All Nitrogen Oxides
CO	Carbon Dioxide	O	Oxygen radical (single atom of O)
HC	Hydrocarbon	O_2	Oxygen
H_2S	Hydrogen Sulphide	O_3	Ozone
NO	Nitric Oxide	SO_2	Sulphur Dioxide
NO_2	Nitrogen Dioxide		

INTRODUCTION

'Florence Fumes' was a headline in a recent newspaper, not, as you might think, concerning some well-known actress or television star in a tantrum, but over a story of air pollution from the heavy traffic in the Italian city of Florence. What connection is there between this news item and the Earth Summit, the conference of world leaders in Rio de Janeiro 1992? One connection is OZONE. It featured in both events – in the former as a pollutant gas – in the latter as one topic for discussion on the destruction of the ozone layer. In one case ozone is bad, in the other it is good: no wonder that people are perplexed. 'Ozone Friendly' labels on cans of hairspray or air freshener, only add to the confusion!

Has ozone something to do with the Greenhouse Effect? Well it has ... and it hasn't! This book is an attempt to resolve some of the confusion by explaining the environmental problems associated with ozone. In order to do this it is necessary to recognize that the problem with ozone is different at ground level where we sleep, work and play, from that of the stratosphere some 20 to 30 km above our heads. In the following pages '**down here**' describes the ground level problem and '**up there**' describes that of the stratosphere. Perhaps most important of all, examples are given of the way ozone affects the lives of people in various parts of the world, and the dangers all of us face when we try to escape to enjoy leisure time in the sun. Finally, ideas are suggested of some ways in which each one of us can live in order to improve the situation.

Ozone is but one air pollutant – there are many others. It may be convenient to write a book about just one environmental problem, but do remember, conservation problems are inter-connected – to try to resolve one without considering the others would be extremely stupid.

The year AD 2000 is seen by most of us as a milestone in human history – if by then the nations of the world have not tackled most of the major environmental problems, including that of ozone, the future will look very bleak. Remember that even now all is not doom and gloom – Planet Earth is a wonderful place – by trying to understand our environment better you are already taking a major step forward towards CONSERVATION 2000. I hope that this book will help you.

Sun burn

Sun Smart

If you lived in Australia you would know all about the cartoon character Sid Seagull and his 'Slip Slop Slap' slogan to encourage the prevention of skin cancer from too much exposure to the rays of the sun. He appears on posters, car stickers and television advertisements. His message is: 'Slip on a shirt. Slop on sunscreen. Slap on a hat.' In other parts of the world 'sunscreen' is known as 'sun lotion' or 'sun cream' and is usually best known by a brand name.

The Sun Smart campaign, to give it its proper title, is organized by the Anti-Cancer Council of Victoria, one of Australia's states. Australia is so sunny that it is one of the high risk skin cancer areas of the world. The Council claims that two out of three Australians will develop skin cancer when they are adult, as a result of too much sun on their skins when they were young. It states that too much sun will:

- increase the risk of sun damage and skin cancer
- make you wrinkly before your time
- thicken your skin and make it look like leather
- cause painful sunburn.

ANTI-CANCER COUNCIL OF VICTORIA

Sun smart

In order to prevent this you should:

- stay in the shade between 11 a.m. and 3 p.m.: these are the hours known as the 'UV radiation danger zone', when the sun is at its strongest
- use tree-shade, hat-shade and shirt-shade when outside
- use a sunscreen which will help protect you against skin damage but will allow the body to become tanned.

The message is to 'treat the sun with a healthy respect'.

One thing is emphasized: sometimes it may feel less hot – for instance on a windy day – but the sun's rays are still damaging to the exposed skin. The ozone layer shields the world from the worst of these rays, so any loss of protection because of a thinner layer above, is of vital importance in Australia.

The measurement of ozone

In the USA and the UK, the amount of ozone in the air is measured as parts per billion (ppb). Traditionally the UK billion is a million million whilst that for the USA is a thousand million. Since the American unit is taken for use by scientists, it is best to use the measurement **pptm (parts per thousand million)** in order to avoid any confusion. To com-plicate matters even further, in many other parts of the world, e.g. Germany and Switzerland, concentrations of ozone are measured as micrograms per cubic metre (pg/m^3). To convert pptm to pg/m^3 multiply by two. Thus 100 pptm is the same as 200 pg/m^3. In this book the measurement from now on will be shown as **pptm**.

SunSmart

Friends of the Earth carried out a survey of NO_2 in several major towns in England during a four-month period over Christmas 1992. The results revealed that at five sites in London, two sites in Manchester and one site each in Birmingham and Cardiff, the European Community safety levels for NO_2 were exceeded. People using the very busy City Road in London were breathing air with average readings of 59 parts of NO_2 per thousand million parts of air. An average means that there must be higher and lower levels during the month. So on some days the air must have been highly polluted – any sunshine would soon cause the nitrogen gas to create O_3. The Government's Environmental Minister dismissed the study as 'cheap and cheerful', but on the same day as the survey was published the same minister officially opened the eighteenth air quality monitoring station in central London.

Scenario I

Date October 1991

Location Athens, capital city of Greece

Weather For a week there were unusually high temperatures for autumn, rising to 95°F (35°C) – sunny with clear skies.

Special circumstances Athens has a million cars using its crowded streets. For traffic control purposes the city is divided into an outer and an inner zone. For ten years there have been controls to prevent the occurrence of smog. A system of car restrictions has been in force where only cars with even registration numbers have been allowed into the inner zone on one day with those with odd numbers on the next – and so on. The number of taxis has also been controlled. Athens is situated in a bowl-shaped area of land with hills all around it. This contributes to the problem, as temperature inversion, a layer of warmer air above the less warm air at ground level, prevents the smog from escaping.

What happened During the first two days of October and again, two weeks later on 16 October, the build-up of smog reached a peak as the sun shone and the temperature soared. All private cars were banned from the inner zone and only half the taxis were allowed to operate. In the outer zone a ban on half the private registered cars was imposed. Despite these precautions, the nitrogen dioxide levels reached a record 350 ppm, a staggering 1,400 times above the safety limit. Ozone was generated to a level of 245 ppm, yet the World Health Organization (WHO) recognizes a safe level to be between 76 and 100 pptm. This means the danger limit was exceeded by between 3,250 and 2,500 times; an enormous excess creating a very dangerous situation.

Over 300 people were admitted to hospitals with heart and breathing problems. Many of these died over the next few weeks. From the main streets the usual magnificent views of the Acropolis with its Parthenon and other temples, Hadrian's Gate and the Temple of Zeus, were only vague shapes as they were hidden in a haze of thick air pollution. Pedestrians, motorcyclists and cyclists wore smog masks. It is not known what the long term effects will be on the health of Athenians. As one of them said, it was 'like being in a burning haze inside a giant steam cooker'.

(The Athens scenario is based on newspaper reports of 2 October 1991 by Helena Smith (*The Guardian*) and Paul Anast (*Daily Telegraph*).)

From an almost deserted main shopping street usually crowded with tourists and Athenians the Acropolis is hidden from view by the NEFOS (Greek for cloud), the smog layer smothering the city. Tiredness, a feeling of heaviness in arms and legs, sore eyes, coughing and sneezing, vomiting and violent headaches are but some of the minor discomforts suffered by people during a Nefos incident. The solutions: (1) A drastic cut-back in traffic or a change to non-polluting transport. (2) Creating more green park areas. (3) Planting more trees.

Scenario II

Date July 1990
Location Bristol, S.W. England
Survey Site Blaise Castle House, northern outskirts of city. 1 mile south of M4 motorway.
Weather Sunny and clear: wind blowing from N.E.
Special circumstances The beginning of most English schools' long summer break saw the start of family holidays and led to an increase in traffic on the M4 and M5 motorways which feed the summer resort towns and other holiday areas. Inevitable delays on the River Severn Bridge, the road crossing from England into Wales, caused extensive traffic tailbacks of stationary or very slow-moving vehicles with their engines idling.
Result Ozone build-up above the maximum WHO limits.
Conclusion The vehicles in the traffic jam produced excessive exhaust emissions which were blown to the S.W. from the N.E. over Bristol. The fumes combined with the strong sunlight to produce excess ozone.

The City of Bristol established its own monitoring station in July 1989; during the first year of its operation the WHO maximum guidelines were exceeded on five occasions. Bristol is located immediately to the east of the M5 motorway, and has a high density of motor vehicles. The usual prevailing westerly winds carry pollutants across the city.

(Based on the report by David Muir, Senior Scientific Officer, Environmental Health Department, Bristol City Council, 'London Environmental Bulletin', Winter 1990/91.)

Around the world

Although Los Angeles (page 30) and Athens are plagued with ozone, and Bristol too, when the conditions are 'right', it would be unfair to suggest that they are much different from other large towns around the world. In Germany one writer based in Tübingen described a typical local scene: 'The trees die, the traffic jams last for hours: in a muggy, mild winter the smog never seems to have cleared' (Christopher Harvie, *Daily Telegraph*, 25 January 1992). In Italy another journalist wrote:

'Italy's cities have become nightmares of traffic congestion, locked into foul-tempered immobility for hours as cars barge their way into the centres and along ring roads, jostling, hooting and crashing, especially during peak hours. In Milan, Rome, Turin and Naples, pollution has reached dangerous levels, obliging children to go to school wearing industrial masks.'

Mexico City is another town suffocated by air pollution. In November 1991 the situation was so bad 26,000 taxis were banned from the city streets for one day a week in a desperate attempt to beat pollution. By the start of 1992 it was worse. A Metropolitan Commission for Control and Prevention of Environmental Pollution was set up which immediately ordered industry in Mexico City to cut output by three-quarters.

It is only the lack of space which prevents examples of traffic polluted cities in Japan, China, Russia and other 'developed' countries from being given.

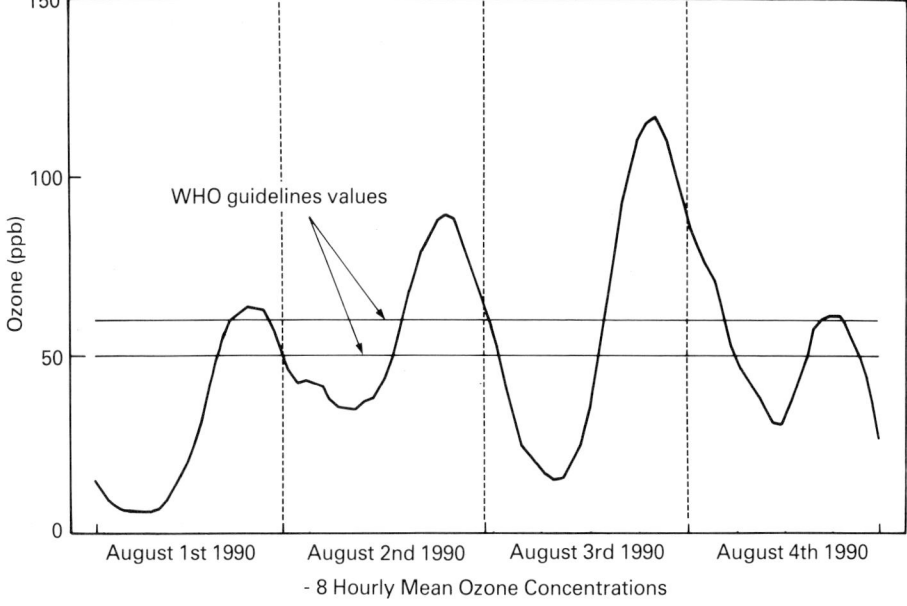

- 8 Hourly Mean Ozone Concentrations

The levels of O$_3$ at the Blaise Castle monitoring site in Bristol in July 1990 exceeded the WHO danger levels on four successive days during the late afternoon.

WHAT IS OZONE?

What ozone is not:

'The seaside is so bracing – breathe in the ozone.' This is a common statement when people are on holiday beside the sea. But the fresh salty air is **not** ozone – it's 'fresh salty air!' One of the most famous of all advertisements for a holiday at the seaside made this mistake.

SKEGNESS

Ozone has already been mentioned several times, but in order to understand the problems that it causes, it is important to know about the chemical processes by which it is formed. Before starting, remember that:

• An ATOM is the smallest particle existing normally in nature although it can be split into other parts.
• A MOLECULE consists of several atoms of the same or different elements to form various chemical substances.
• The chemical symbol for ONE atom of oxygen is O.
• The chemical symbol for TWO atoms of oxygen is O_2 – called Diatomic oxygen.
• The chemical symbol for THREE atoms of oxygen is O_3 – called Triatomic oxygen or more commonly known as OZONE.

Ozone is a gas, each molecule of which has three oxygen atoms: it is a form of oxygen. The gas we breathe to keep us alive is OXYGEN with TWO atoms of oxygen, so it is easy to see that if this has another ONE atom of oxygen added to it we get OZONE. From now on unless I say differently, or use the chemical symbol O, when I say OXYGEN I shall be referring to the oxygen gas that is our essential life support. Oxygen is very stable, that is, it takes a lot of energy to break it up into two single atoms. Ozone is not very stable and can readily change back into oxygen with a single O atom left over – or it can change into some other chemical form.

• Oxygen is a colourless and odourless gas.
• Ozone has a pungent smell and is a light blue colour.

Without oxygen, there would be no life on Planet Earth – it is vital to us. Without ozone, life on Planet Earth would be BOTH better and worse. Why? . . . the answer comes later.

FRANK NEWBOULD after J HASSALL

The chemistry of ozone

Ozone consists of three separate atoms joined together to form one molecule: $O + O + O = O_3$. In nature this is normally caused when the sun is shining on the gases in the atmosphere around the earth. This is called a PHOTO-CHEMICAL reaction. What happens is this. Sunlight acts upon oxygen and splits the molecule into two single atoms: $O_2 +$ **sunlight becomes $O + O$**. Other oxygen then collects up the O atoms to become ozone: but this will only happen if a molecule of any other substance is present. Nothing happens to the other substance – it just causes the chemical change to take place. This is known as a CATALYST.

$O_2 + O +$ **any other molecule becomes $O_3 +$ that other molecule**. But ozone is very unstable, and soon changes again. This is called DECOMPOSITION and again it needs sunlight to make it happen: **O_3 with sunlight becomes $O_2 + O$**. It also happens that ozone and the single atoms react together to become two molecules of oxygen: **$O + O_3$ becomes $O_2 + O_2$**. As long as sunlight is present, these changes cause oxygen to become ozone while at the same time, ozone, molecules decompose back into oxygen. Sunlight is absolutely essential – no sun, no formation of ozone. Therefore no ozone is formed at night, and little ozone is formed below the clouds in daylight, when most of the sun is blocked. There are other chemical reactions which can cause ozone to be created involving gases other than oxygen – more of this when we come to human actions in the ozone story.

Ozone is not a common gas, but it does make its presence felt. It is very important for several reasons, but its main chemical property is that it is very unstable. There is a proverb which says, 'Two's company – three's a crowd'. This may help to fix in our minds the most important point, that oxygen stays together but ozone breaks up!

How scarce is ozone normally?

Scientists talk about 'ppm' or 'ppb' (pptm) (see box on page 7) – of gases making up the air. In other words, 10 ppm means 10 parts of the gas for every million parts of air (not very much) and 10 pptm means 10 parts of gas for every thousand million parts of air (even less). Ozone is normally in the atmosphere at very low concentrations of between 20 pptm and 30 pptm. WHO suggests that between 76 and 100 pptm of ozone is a safe amount to which people can be exposed for a period of an hour. For a longer eight-hour period, the amount should be no more than between 50 and 60 pptm.

Ozone Terrace in Lyme Regis, reminds us of how salty seaside air was once mistaken for ozone.

Above the clouds

One of the best things about daytime flying is that whatever the cloud cover when you take off from the airport, directly the plane breaks through the mist of the cloud outside the windows, it flies into bright sunshine, with the clouds, now underneath, looking like a snow-covered landscape. Although the air through which the plane is flying is 'rarer' – it has less of the gases in it compared with air at ground level – there is still sufficient oxygen in it for the sunlight to cause photochemical changes. This means that while this part of the earth is in daylight, ozone is formed and decomposed all the time. High in the atmosphere, especially in the stratosphere (see page 40) between 20 and 25 km above the ground, ozone is fairly common, although even here, where it is at its densest concentration there are usually only 10 ppm.

Ozone occurs in the stratosphere – it also occurs **down here** nearer the ground, in the troposphere. This is the ozone we inhale, how much depends on the weather conditions. Obviously sunny, cloudless skies can lead to more ozone being formed than in places where clouds are more usual and the sunshine less. Certain parts of the world have more than their 'fair' share of ozone.

To summarize, ozone is an oxygen gas which breaks up and forms very easily in the presence of sunlight. It is slightly blue in colour and has a pungent smell. It is to be found **up there** and **down here** . . . and what a difference where it is found makes to human beings, other creatures and plant life. We will see how it affects us in each case, in the following chapters.

With an artist's eye

My cousin is a painter of landscape pictures. He really is very good and has had several exhibitions of his work. We were admiring the view of the local hills recently when he asked, 'Can you see the blueness of the distant air? Over the years I have noticed how much the clarity has changed – have you any idea why?' Extra ozone perhaps!

Oxygen is vital to our health. Extra oxygen is supplied for emergency use in aircraft. Here, a Royal Navy Sea King pilot carries an air pump and oxygen equipment.

DOWN HERE – EFFECTS ON LIFE

Robert Louis Stevenson wrote a book about Dr Jekyll, who, after his experiments with a 'magic' potion, changed into Mr Hyde. Dr Jekyll became a dual personality. The good side of his character was offset by the dastardly behaviour of his other self Mr Hyde. Ozone has a dual personality, too. **Up there** it is helpful and useful – **down here**, it can be harmful and injurious to people.

Down here

What effect does ozone have **down here**, at ground level? Above the minimum safety level of about 40 pptm, all those who inhale it or come in contact with it may have:

- lung irritation
- headaches
- aching throats
- a feeling on the edge of being really ill
- general nausea, maybe even vomiting
- soreness to the eyes

These trees in Czechoslovakia have been killed by ozone attack, among other factors causing disease.

For those unfortunate people who normally suffer from bronchial trouble which makes breathing difficult, an increase of ozone in the air they breathe will aggravate the original condition. At times when the concentration of ozone is high, the incidence of chest infections will increase. Where the sufferers are acute patients (they are very ill), death may occur unless immediate treatment is available.

But it is not only human beings who are affected. Other animals and plants show a reaction to ozone, so much so that farmers may suffer financial losses, estimated to run into millions of pounds or dollars.

In July 1989 the Office of Technology Assessment (OTA) in the USA, reported that studies of the effects of ozone on animals showed that the gas altered the structure and the chemistry of their lungs. If similar results occurred in humans the effect would be very serious indeed. In 1989, also, the American Lung Association (ALA) reported experiments which had been carried out with ten men. These men were non-smokers so the results could not be blamed on the fact that they inhaled tobacco smoke. For over six hours each man was made to inhale air which first of all contained 80 pptm of ozone and for another six-hour period, 100 pptm. What happened?

1 Even at 80 pptm the cells in the lungs which destroy harmful bacteria were reduced sufficiently to allow infection to get into the body more easily. The cells are known as macrophages.
2 At 80 pptm the body cells and chemicals which cause redness and swelling to occur (inflammation) increase a great deal. Extra white blood cells are an example of this.
3 At 100 pptm the levels of protein in the lung increased as did the level of a substance called Prostglandin E2. This E2 is an important controller of the immune system of the body; it is the body substance which stops all the bacteria and viruses we breathe, touch, drink or eat from causing us illness every time.

Other studies have shown that increased levels of ozone not only affect lung disorders (asthma, bronchitis, emphysema, chronic

---

The effects of ozone on human health

In Switzerland an organization called the Physicians for Environmental Protection carried out many tests to discover what would happen to people of different ages when exposed to various levels of ozone for different lengths of time. These were the main results:

Amount of ozone	Length of time	Effect
60 pptm	$\frac{1}{2}$ hour of hard work	The mucous membranes of the eyes, nose and throat become irritated. People found they were unable to play sports so well.
120 pptm	$\frac{1}{2}$ hour of hard work	All the first symptoms but to a far worse degree.
120 pptm	2 hour's work	The lungs of children became less efficient – they were 'out of breath'.
150 pptm	1 hour's work	Adults began to suffer loss of lung power and suffered severe eye irritation.
200 pptm	3 hour's exposure	People were unable to adjust their eyes to darkness.
250 pptm	Highest level during the course of one day	Coughing and pains in the chest whilst working normally.

(Source 'Air Pollution and Health' quoted in *Acid News* January 1991)

airway obstruction disease) but also illnesses associated with the heart and blood circulation.

As with many disorders, increased levels of ozone affect babies of up to two years of age, more than older children. Also at risk are elderly people and pregnant women. Obviously people already suffering from any of the diseases mentioned will have them made worse with exposure to ozone-polluted air.

Effect on vegetation

The effect of ozone on plants has been investigated too. Experiments have shown that photosynthesis has been reduced in plants exposed to levels of ozone as low as 50 pptm if this has been present for more than 16 days during the growing season. For 100 pptm it takes only ten days to cause damage and if the level is as high as 300 pptm then damage is done in only six days. Leaves of plants become discoloured, a fact which has been used in some ozone level investigations as we shall see shortly with the WATCH surveys carried out by children in the UK (page 29). Other symptoms include the loss of leaves, especially from trees. Plants grow more slowly in the presence of ozone and their resistance to pest attack becomes less. As with people who have an illness already, it is 'easier' to pick up other diseases when exposed to air pollution. In Germany and parts of the USA, near Los Angeles for instance, estimates suggest that 85 per cent of pine

tree losses are due to ozone pollution.

Crop yields are reduced dramatically if ozone levels are increased up to 900 pptm. Maize (corn) loses up to 13 per cent of its expected yield, soya beans as much as a third and wheat nearly a quarter. One estimate suggests that 48 million tonnes of grain were lost worldwide in 1987 due to ozone pollution. Worse is to come when we consider the effect of reduced ozone **up there**.

Cross-section of a leaf. The stomata are pores on the underside of leaves, which open and close to allow gases in and oxygen out. Ozone closes these pores and causes the leaf to wilt with damage to leaf cells.

Leaf destruction – why trees die

Tests have shown that ozone damages the cells, walls and other structures of leaves. Where trees are concerned the breakdown of these parts of the leaf allows essential magnesium to leak out and to be washed away by rainfall. The trees cannot replace the magnesium rapidly enough so that it suffers increasing magnesium deficiency and death results. Trees at higher altitudes above the mists, where the sun shines for longer, seem more likely to suffer ozone damage than those situated lower down.

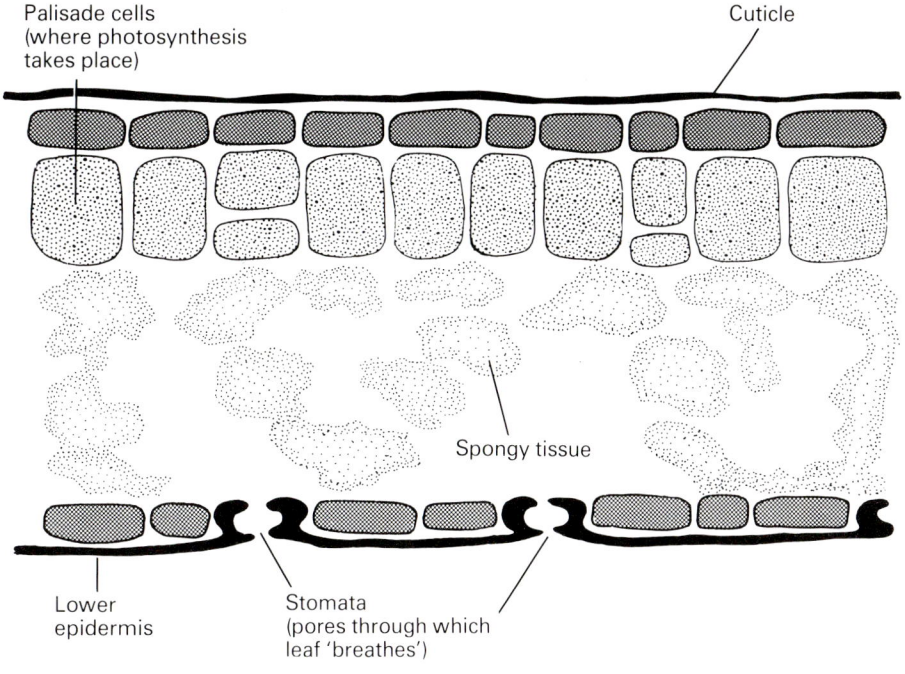

Palisade cells (where photosynthesis takes place)

Cuticle

Spongy tissue

Lower epidermis

Stomata (pores through which leaf 'breathes')

All the colours of the rainbow

Most of us have seen a rainbow as sunlight passes through rain-drops. This is the *spectrum* and is caused by the white invisible light breaking up into its many parts all of which vary slightly in wavelength. In other words, white light consists of seven basic colours and when it is dispersed it breaks up into its various parts. Red is found at one end, where the wavelength of the rays is the longest and a violet colour occurs at the other end, where the wavelength is shortest. As the wavelength of the rays becomes longer, the sun's radiation is invisible and because they are below the red end of the spectrum they are called 'infra-red rays' and are associated with heating. Beyond the shorter end of the spectrum where the colour is violet, the rays are also invisible and are called 'ultraviolet rays'. These rays are sometimes known as *black light*. (In Latin the prefix *ultra* means 'beyond' and *infra* means 'below'.) Wavelengths are measured in metres but those of the sun's radiation are so small that it is usual to use Angstrom units (Å) where one of these units is equal to 1/10,000,000,000 of a metre. Ultra-violet radiation has wavelengths of between about 1,000 Å and 4,000 Å and is divided into UVA rays under 4,000 Å, UVB rays between 3,150 and 2,800 Å and UVC rays 2,800 to 1,000 Å. The UVA is absorbed by oxygen and most of the UVB and all of the shorter wave UVC are removed by ozone.

X-rays are short waves known for their ability to make certain parts of the body show up on special photographs which we know as X-ray pictures. We also know that they are dangerous, so lead shields have to be used to prevent damage being done in places where the rays are not wanted. The reason there is so much concern about UV rays reaching the earth is because they, too, are a danger to human bodies, especially the UVB rays. The UVC rays, with their shorter wavelength, would be lethal to us unless we wore some sort of 'space suit' protection.

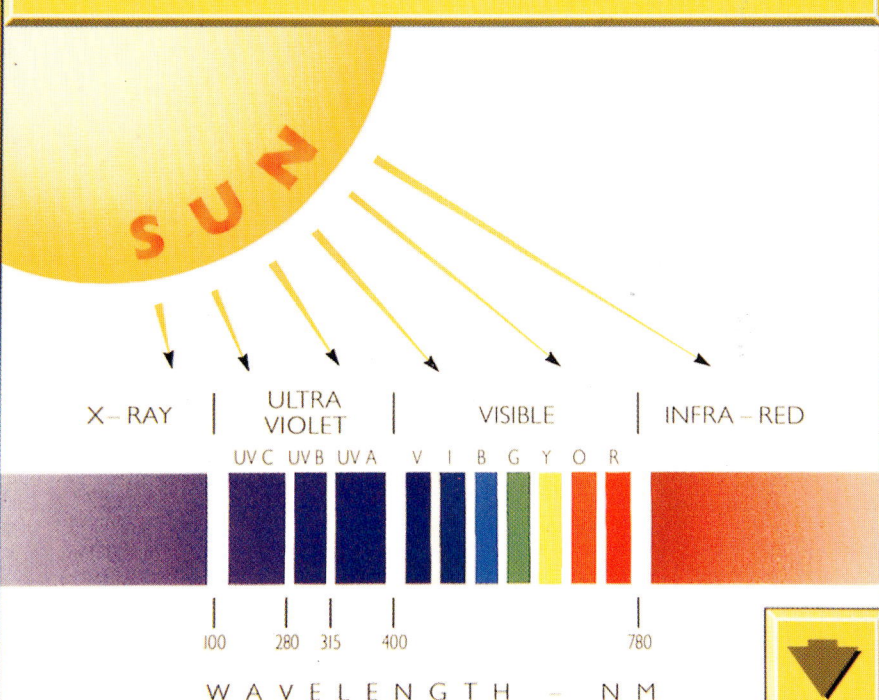

THE ELECTRO-MAGNETIC SPECTRUM

SUN

X – RAY | ULTRA VIOLET | VISIBLE | INFRA – RED

UVC UVB UVA   V I B G Y O R

100   280 315   400   780

WAVELENGTH – NM

The diagram shows the electromagnetic spectrum stretching from the long infra-red waves to the short ultraviolet and X-rays. Not indicated are the very long radio and micro-waves and the extremely short gamma rays. The visible light rays which make up the rainbow are sandwiched between the infra-red and ultraviolet rays. The wavelengths are shown in metric measurement (see page 18).

Photochemical change

Before we learn why it is that ozone has a friendly 'Dr Jekyll' side to its nature **up there** in the stratosphere, we need to know something more about the processes of nature. I have already said that with sunlight, oxygen becomes two single atoms, and that ozone with sunlight decomposes into oxygen and ozone. Sunlight is essential to the changes which take place. In causing these changes the sun's rays are turned into heat and thus are lost in their journey to the surface of the earth. If they were not intercepted by the oxygen or ozone the rays would pass through the stratosphere and so reach the part where we live with various results, as we shall see. Scientists call this reaction of the sun's rays with gases in the air PHOTOCHEMICAL CHANGE.

Photochemical changes occur anywhere that light reacts with a chemical. For the moment we are concerned with it happening in the layer of atmosphere between 20 and 25 km above our heads, but it also happens in the ozone produced at ground level . . . definitely not to our advantage. Ozone is only a minor gas in the make-up of the atmosphere, to be found in different concentrations between ground level and the level where the atmosphere ceases to exist in any real degree, that is, about 300 km high. The majority of the ozone is found in the middle stratosphere, but even here, only about one molecule in about 100,000 is ozone (only 0.001 per cent). Yet this is enough to remove most of the UVB rays, and all of the UVC rays, the ones which do us harm. Another measure which shows how little ozone there is **up there** is the fact that of all the O atoms in the air of the stratosphere, there are 50 million times as many of oxygen as there are of ozone. This is both a very difficult fact to estimate and to understand. Another calculation says that if all of the ozone in the upper atmosphere was compressd into a band all around the earth like an orange skin, it would only be as thick as a piece of window glass or the thickness of a £1 or silver dollar coin.

Usually the content of the atmosphere is said to be in equilibrium – this means that the amount of ozone that is lost is balanced by the amount which is created so that, in total, it stays about the same . . . at least it did!

American ER2 research planes, are used to investigate pollution at high altitudes (see page 28).

Alpha, beta, gamma are the first three letters of the Greek alphabet, that is, they are A,B,C. These letters are used for the various wavelengths of ultraviolet radiation UVA, UVB, UVC. We usually refer to the UV rays as UV-alpha, UV-beta or UV-gamma rays. Don't be confused by the very short wavelength waves called gamma rays, which are not connected with ultraviolet.

UV destruction by ozone

Molecules of ozone within the atmosphere, some 12 to 50 km above the surface of the earth, are in the path of the UV rays emitted from the sun travelling towards the ground. Most are concentrated in the 20 to 25 km region. As already explained, ozone molecules are unstable and quite ready to shed one of the atoms to make oxygen and O (page 12). When the UV ray strikes the ozone's three atoms, separation occurs, with the energy of the life-damaging ultraviolet being changed into heat, harmless enough at this height, and quickly lost into the surrounding rarefied air. Thus the possibility of the UV rays arriving at ground level is reduced as one molecule of ozone after another intercepts the 'enemy', rather like fighter planes preventing bombers reaching their target. Each break-up leaves a free oxygen atom (O), and an oxygen molecule floating in the air, each prepared to rejoin each

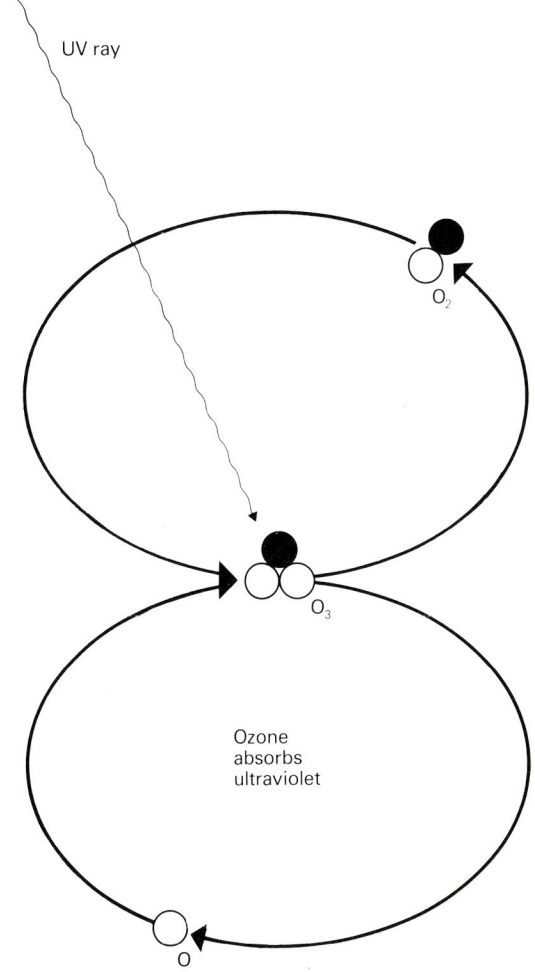

other or link up with other floating oxygen as soon as possible. They recombine to form new ozone ready to get in the way of another UV ray – and so it goes on – we could call it an *ongoing cycle*.

In this way the intense bombardment of the dangerous ultraviolet radiation from the sun is rendered harmless, even though the concentration of ozone is extremely small. This photochemical process is essential to the well-being of life on Planet Earth even though only 0.001 per cent of the atmosphere consists of ozone at its maximum density. What is most important is that the state of equilibrium is maintained – that the amount of ozone

O_3 absorbs UV rays. UV rays break up an O_3 molecule which splits into O_2 and O and which reunite to form O_3 (left) when the cycle is repeated. The UV rays are dissipated in heat.

destroyed is balanced by the amount of ozone created, a situation that is threatened by the activities of people. It is as if the anti-aircraft fire from the ground is destroying our own friendly fighters rather than the enemy bombers. What we are sending up into the atmosphere is doing the damage! Destruction is taking place, in the words of wartime generals, by 'friendly fire'. Equilibrium is NOT being kept.

UP THERE – EFFECTS ON LIFE

Where the ozone layer is thin, or worse still, missing altogether, dangerous rays from the sun will reach the earth in above normal amounts. Human skin contains cells of melanin, a pigment which gives the external body colour. The more melanin there is, the browner the colour of the skin. Melanin acts as a barrier to UV rays. Thus 'white' skinned people are more likely to have damage done to their skin than those who have brown skins. Normally this would be in places where the sun is strongest, the tropical or sub-tropical areas of the world. Ozone loss **up there** means other parts of the world are now places where the sun's action can affect those with less melanin than others. An increase in the occurrence of various types of skin cancers is one result of ozone depletion in the stratosphere.

Experiments have shown that an increase in UVB radiation will stunt the growth of plants and reduce the amount of food which is produced from them. Any plant that gets too much UVB will die. These plants include those which live in the sea, the phytoplankton, on which the whole of the marine life ultimately depends, and whose existence partially controls global warming. Zooplankton, too, will be badly affected: zooplankton is made up of a myriad of small animals which 'graze' on the phytoplankton and provide food for larger sea animals, especially fish. Land animals are unable to resist increased levels of radiation. Much depends on the areas in which they live (they are more protected in forests than on plains) and their sort of skin covering. An increase in ultra-violet radiation, due to the depletion of the ozone layer, poses a serious threat to the welfare of all life on Planet Earth.

The effect on health

Less ozone **up there** means more UVB **down here**. Ultraviolet B radiation stimulates the growth of carcinomas, the name given to those forms of cancer which arise in the skin and the internal organs of the body (sarcomas are the kind which occur in muscles, bone, blood vessels, fat and lymph glands). Three sorts of carcinomas affect the skin: they are called 'basal cell carcinomas', 'squamous cell carcinomas' and 'melanomas'. The first are called 'basal cell' because they arise in the basal cell layer of the epidermis. They usually occur on the

This is a picture of a melanoma skin cancer discovered before it was too difficult to treat.

face as small, round or flattened lumps. The only advantage of these sort of growths is that they rarely spread to other parts of the body, and, if treated early, can be cured. The second sort are the 'squamous cell carcinomas' which are less common but more dangerous. They appear on the hands, forearms, face and neck. They usually have scaling red areas which may bleed easily and ulcerate or look like unhealing sores. They may spread to other parts of the body. The third sort, the 'melanomas', arise in the cells of the skin which contain melanin. They often develop from a mole or a wart already on the skin. Unfortunately melanomas grow quickly and spread to other parts of the body, especially to the liver. They need to be treated as soon as they are recognized. This is not the book to discuss medical matters in detail, but it is worth noting that anyone who has a wart, mole or other blemish on their skin which starts to grow, itch or bleed should have it examined by a doctor. It is the purpose of this book to relate the dangers of the loss of ozone **up there** with the extra chances of developing skin cancer **down here**. Health precautions will be mentioned later in the book but here it is important to stress that sunbathing can be dangerous to people with light-coloured skins. If you are not used to exposing your body to strong sunlight it is essential to sunbathe for short periods only, having used barrier creams on the skin.

Another form of damage to the skin by extra UVB rays is not particularly dangerous but it is something which people wish to avoid as far as possible. Wrinkling of the skin comes with old age: indeed the term 'Wrinkly' is an expression sometimes given to the elderly. Wrinkling can be hastened by exposure to sunshine and thus may begin even earlier with exposure to more UV rays.

The clouding of the lens of the eye is called a cataract. Because it can cover the lens altogether, like a white sheet over a light bulb, it will prevent clear vision. It can be removed and sight can usually be restored. The clouding is caused by a thickening of the proteins in the lens which can be stimulated to grow by UVB radiation. Most cataracts are caused by other means especially with older

Even those people with dark skins who live in high sunshine areas of the world, become wrinkled at a relatively early age.

people but ozone depletion, and the extra radiation it lets onto the earth's surface may mean an increase in this disorder among younger people.

Some scientists have tried to predict the amount by which health disorders will increase with various changing levels of ozone in the stratosphere. For example one group of experts predicts that a six per cent decrease will lead to a rise of over a third in basal cell cancers by the year 2030, and that for every one per cent reduction in protective ozone, an extra 100,000 people will develop cataracts. These estimates can only be scientific-ally-based guesses, but even if they prove to be inaccurate, the

basic message is clear: 'extra UVB is dangerous to human health — so we must stop the depletion of ozone.'

The effect on other living creatures

Any increase in the amount of UVB striking the earth will have a serious effect on living creatures of all kinds. In particular, small, young creatures and plants, will be affected. The many varieties of the very small creatures which make up the zooplankton will be harmed. Since these form the

food of larger marine life, which, in turn, provide the food for the fish caught for human consumption, the vital maritime food web will be broken.

The effect on plants

In plants, including food crops such as wheat, rice, soya beans, barley, potatoes and beans, growth and germination (seed-making) will affect productivity. A decline in the process of photosynthesis will be one of the mechanisms by which the extra UVB will adversely affect plant growth. In experiments carried out by Australian scientists of the Commonwealth Scientific and Industrial Research Organization (CSIRO) there was a loss in pea plants of over half of the chlorophyll (the green-coloured matter in the leaves of plants). Chlorophyll is absolutely essential to plant growth, so the loss of harvest is an obvious outcome of an increase in UVB. Another plant essential is a protein called rubisco which allows leaves to 'breathe' carbon dioxide. The CSIRO scientists recorded losses of over a quarter of the rubisco in plants subjected to extra UVB.

Trees will not escape the damaging rays. UVB is especially injurious to trees which have been weakened by some other form of attack, whether fungi, insects, drought or acid rain. But of all the plant life on earth, it is probably the plants we do not see in our normal life – the phytoplankton – which will be most affected. Scientists from Texas University, USA, have

SKIN TYPE	MODERATE UK, N. Europe		HOT Mediterranean		VERY HOT Tropics, Africa	
	First Days	Following Days	First Days	Following Days	First Days	Following Days
CHILDREN and SUN SENSITIVE SKIN, (fair or red hair, pale skin and freckles)	10 to 8	8 to 6	15 to 10	10 to 8	25 to 20	15 to 10
FAIR SKIN, BURNS EASILY (fair hair, blue eyes, mid to pale skin tone)	8 to 6	6 to 4	10 to 8	8 to 6	15 to 10	10 to 8
TANS NORMALLY, TENDENCY TO BURN (dark hair and eyes, mid to pale skin tone)	6 to 4	4 to 2	8 to 6	6 to 4	10 to 8	8 to 6
TANS EASILY (dark hair and eyes, olive skin tone)	4 to 2	2+	6 to 4	4 to 2	8 to 6	6 to 4

found that a ten per cent increase in UVB rays will kill almost all of these small plants floating in the oceans. Even a one per cent increase will slow photosynthesis and badly affect their growth. Phytoplankton is the food supply upon which all marine life depends – and these myriads of plants absorb carbon dioxide from the air and so make an enormous contribution to reducing global warming.

An important factor where plants are concerned is the enormous range of varieties which occur. This is known as biodiversity. UVB will attack each species to a greater or lesser extent. This means that it is quite likely that extra harmful rays will result in a serious reduction in biodiversity which will have damaging effects on the whole natural balance of a large area.

Each sunscreen is given a Sun Protection Factor (SPF) number, which indicates the approximate extra length of time it will allow you to stay in the sun without burning. The higher the number, the greater the protection so that a SPF of 4 will allow you to be exposed to the rays of the sun for 4 times as long as without protection. Some sunscreens are called 'broad spectrum' sunscreens. This means that they protect the wearer from UVA and UVB rays.

In 1987 the US Environment Protection Agency (EPA) forecast that over the next 90 years the USA could expect an extra three-quarters of a million deaths, 12 million eye cataracts and 40 million skin cancers as a result of O_3 depletion.

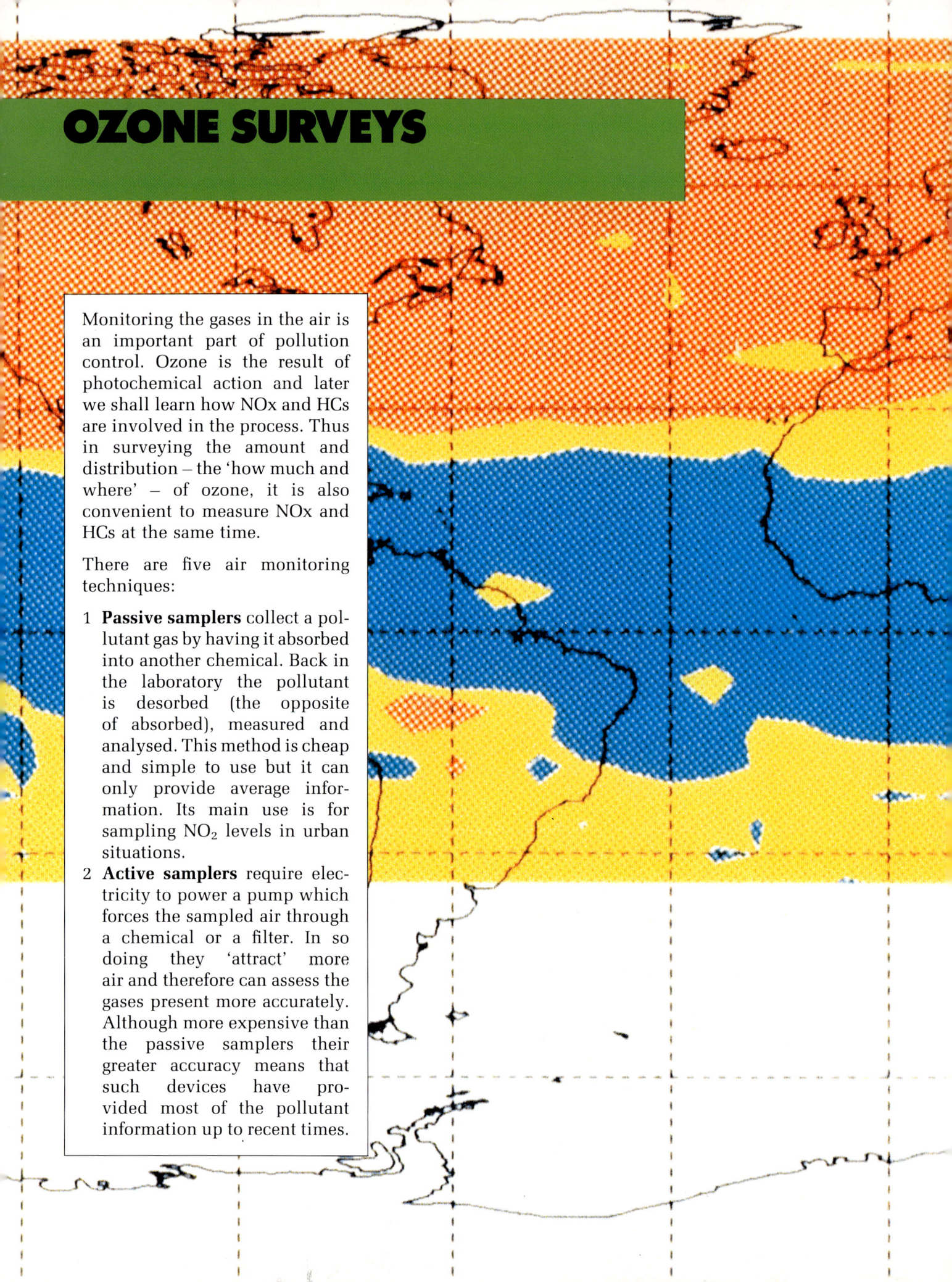

OZONE SURVEYS

Monitoring the gases in the air is an important part of pollution control. Ozone is the result of photochemical action and later we shall learn how NOx and HCs are involved in the process. Thus in surveying the amount and distribution – the 'how much and where' – of ozone, it is also convenient to measure NOx and HCs at the same time.

There are five air monitoring techniques:

1 **Passive samplers** collect a pollutant gas by having it absorbed into another chemical. Back in the laboratory the pollutant is desorbed (the opposite of absorbed), measured and analysed. This method is cheap and simple to use but it can only provide average information. Its main use is for sampling NO_2 levels in urban situations.

2 **Active samplers** require electricity to power a pump which forces the sampled air through a chemical or a filter. In so doing they 'attract' more air and therefore can assess the gases present more accurately. Although more expensive than the passive samplers their greater accuracy means that such devices have provided most of the pollutant information up to recent times.

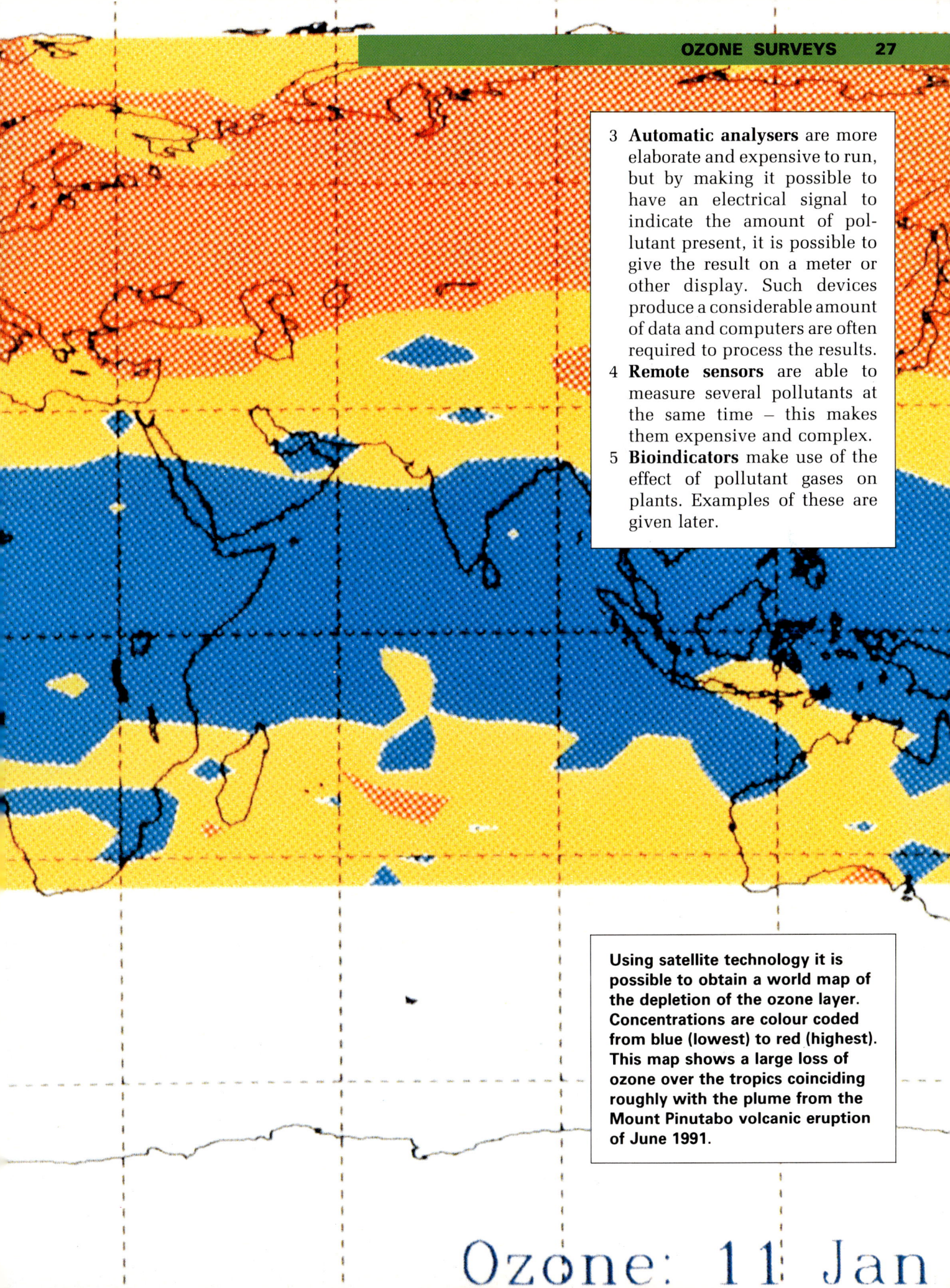

3 **Automatic analysers** are more elaborate and expensive to run, but by making it possible to have an electrical signal to indicate the amount of pollutant present, it is possible to give the result on a meter or other display. Such devices produce a considerable amount of data and computers are often required to process the results.

4 **Remote sensors** are able to measure several pollutants at the same time – this makes them expensive and complex.

5 **Bioindicators** make use of the effect of pollutant gases on plants. Examples of these are given later.

Using satellite technology it is possible to obtain a world map of the depletion of the ozone layer. Concentrations are colour coded from blue (lowest) to red (highest). This map shows a large loss of ozone over the tropics coinciding roughly with the plume from the Mount Pinutabo volcanic eruption of June 1991.

Ozone: 11 Jan

Ozone can be measured using any of these five techniques, but it is the use of the UV absorption analyser which is the most useful. With this, the sample of air passes through a tube and a UV detector determines how much ozone is present.

High level surveys

Up there in the ozone-rich layer, it is far more difficult to estimate the amount of ozone and pollutant gases present than it is to identify and measure them **down here** at ground level. The usual method is to send meteorological balloons into the atmosphere carrying measuring instruments whose readings are transmitted to the tracking station by radio. The name given to this whole system is 'Radio Sonde'.

High-flying planes adapted from the American U2 'spy' planes, have been used to collect data from the upper atmosphere for many years. These civilian versions of the U2, named ER-2, were used to check on the British Antarctic Survey's claim that a 'hole' existed in the ozone layer above the South Pole. In October 1987 these investigations were at their busiest. Since that time NASA's Jet Propulsion Laboratory in California has continued to study ozone levels over various parts of the earth. For instance in November 1991, three ER-2 aircraft were detailed to begin a five-month mission to search for evidence of an ozone hole in the northern hemisphere. These planes fly at altitudes of up to 70,000 feet (about 20 km) above the earth's surface. In particular the high polar clouds were investigated for evidence of pollutant content.

Satellite investigation

Satellites circling high above the earth are another form of transport for the instruments needed to study the very complex chemical processes which destroy ozone. At the same time the instruments on the most recent satellite, known as the Upper Air Research Satellite and controlled by NASA, provide information for climatic modelling where specially programmed computers forecast the effects of global warming. The UARS was launched in September 1991 and among other scientific instruments carried was the Microwave Limb Sounder which measures the amount of ozone, chlorine and water vapour present. It was sent almost 655 km

ELE monitoring station. A gas monitoring unit which can automatically measure various pollutant gases. In 1987, 17 of these stations were set up by the Department of Trade and Industry, with readings taken every four hours.

(400 miles) into space and was intended to operate for two or three years.

Bioindicators

One of the indirect ways of determining how much of a pollutant is present in the air is to look at the effect it has had on plants. With acid rain, for example, it is easy to note the presence or absence of black spots on rose leaves. These spots are the result of a fungal attack which the grower will combat with a spray. If chemicals similar to those used in the spray are already in the air because of high levels of pollution the rose leaves will be clear of black spot. Black spots mean there is no pollutant in the air – lack of black spot probably means the air is contaminated with a pollutant such as the sulphuric acid in acid rain. Lichens are plants which dislike acidic air, some species more than others. It is possible for the effect of different levels of acidity on different lichen species to be determined and a measurement scale worked out. An observer can see which lichens are present in an area, and from this, work out the level of acidity in the air. The orange/yellow lichen *Xanthoria* can be seen on walls and roofs where the air is particularly clean.

Ozone has an effect on plants as well. With the tobacco plant it causes white or pale brown spots on the leaves. If you know these plants you will be aware that they have large oval-shaped leaves which are grown to be dried and shredded into tobacco for smokers. There are many differ-

ent species of tobacco plants which are known in botany as *Nicotiana tabacum*. During the early years of the 1990s, 'WATCH', the junior section of the Royal Society for Nature Conservation (RSNC) in the UK, organized ground level ozone surveys using the spotting properties of the *Nicotiana* leaves. Two different tobacco varieties were used – one, *Nicotiana tabacum*, Bel-W3 was used because its leaves are very sensitive to ozone spotting; levels of only 20–30 pptm of ozone cause spotting to occur. The variety Bel-B starts to spot only when the level reaches 90 pptm. This means that different levels of ozone concentration can be observed and it also means that if Bel-B starts to spot at the same time as Bel-W3, early on in the growing period, it is likely that something other than ozone is the cause. By supplying specially cultivated seeds of both plants to WATCH members, together with instructions, information, a spotting density card, data-recording forms and a thermometer, WATCH set up a research investigation into the levels of ozone **down here** all over the UK.

The amount of blotchiness on a tobacco plant leaf indicates the amount of ozone in the air.

The logo of the WATCH project.

THE OZONE PROJECT

AN INVESTIGATION INTO LOW LEVEL OZONE

DOWN HERE – PEOPLE CREATE OZONE

Ozone does occur naturally at ground level in the air we breathe, normally at no more than 30 pptm. Unfortunately people – or, at least, the machines that people use – create ozone by their activities. We have already noted that ozone is formed when sunlight reacts with oxygen, a natural reaction, but it is also formed when sunlight has photochemical effects on exhaust gases from petrol/oil engines. Industrial processes also produce the pollutants which can create ozone.

Domestic pollution

Long hours of sunshine in some areas of the world encourage people who live in towns to spend as much time as they can outside in their gardens or on the beach. Their lifestyles involve the use of machines and equipment which give rise to ozone, thereby forming pollution. For example, in California, many people own houses with large gardens and swimming pools. Food is cooked on barbecues; motorized ride-on cutters have replaced hand-pushed grass mowers; pool pumps and water conditioners perform their tasks; hydrocarbons and nitrogen gases rise from the natural gas fuels which have superseded the traditional charcoal grills; air conditioning and the other paraphernalia of modern affluent homes contribute their portion of pollution. All in all, it is estimated that as much as 30 per cent of Los Angeles smog emanates from domestic pollution – and ozone is lurking in that smog. Rules against car pollution are to be introduced (by 1998 for every 100 cars sold in California, two must be electric and by 2007 there will be a complete ban on diesel and petrol engines) and a ban on barbecues is also planned.

The way in which ozone affects human health and plant growth has already been described. In addition, it is also one of the greenhouse gases, and though it is not so important in global warming as carbon dioxide, it has its own adverse effects. Certainly, it is an addition we could do without. Because ozone is created by the

The haziness of this picture of Los Angeles on a sunny day indicates that the air is smog laden. Temperature inversion (a layer of warm air which traps the pollution) is clearly visible in the picture.

reaction of pollutants which have been made directly in engines or factory processes, it is known as a **secondary pollutant**. The others are **primary pollutants**. In order to reduce its effects it is necessary to reduce or eliminate the primary pollutants which cause it to occur. This is relatively easy – but do we have the will to do it? It costs money to take the necessary precautions.

The internal combustion engine (ICE) is a boon; without it, modern life would be very different, and travel extremely limited. But the name provides the clue to the problem of the ICE – inside each one fuel is burnt (combustion) to cause an explosion which forces the pistons to move and so cause the wheels to turn. Unfortunately burning the fuel makes unwanted pollutants: these are carbon monoxide, nitrogen oxides and hydrocarbons. Water vapour and carbon dioxide also escape from the exhaust pipes and though they are not poisonous for people they do contribute to the Greenhouse Effect – the environmental problem leading to the earth becoming warmer (refer to *The Greenhouse Effect* in this series of books). Carbon monoxide is a dangerous poisonous gas – it is the gas which causes death when suicide victims pipe car exhaust into a closed car. Hydrocarbons – sometimes known as Volatile Organic Compounds – are many in number, of which benzine, propane and ethanol are probably the best known. There are also several nitrogen oxides including NO_2 and NO. For pollution purposes they are lumped together with the chemical signature NOx. It is the NO_2 and the HCs which increase the chances of ozone being formed, with NO acting as a brake. This is explained in the Chemistry Box on page 32.

General estimates for the UK and the USA suggest that about half of all ozone comes from vehicle exhausts and a third from industrial processes. HCs come from other sources as well as engines. Estimates by UK scientists of the Photochemical Oxidants Review Group are:

- 20% from car exhausts
- 6% from petrol evaporation (when filling a car try not to breathe in the fumes!)
- 6% from oil refineries
- 30% from solvent use – i.e. cleaning liquids and some types of paint
- 20% from natural gas leakage (pipelines, extraction wells)
- 6% from industry

Concern is growing about the increase in pollutants from domestic sources.

Ozone and the Greenhouse Effect

Ozone is a gas which will take in the heat escaping from the earth and retain it, thus preventing it radiating back into space. In the lower layers of the atmosphere, the air will increase in temperature, with the land and oceans beneath becoming hotter. This means that ozone in the atmosphere causes global warming, an environmental problem usually presented as the Greenhouse Effect. Although the warming will only be a matter of a few degrees centigrade, the world will be adversely affected. Climates will change. In parts of the USA,

particularly in the Midwest where much of the grain the world needs is grown, it will become drier, with drought leading to the sort of conditions which resulted in the 'dust bowl' of the 1930s. Already starvation is the result of climatic change in the Sahel area of North Africa: even the vital monsoon rains of South-East Asia may not come every year if global warming continues. In addition to these effects it is likely that sea levels will rise causing flooding of coastal areas, where so much of the global

In Britain, the first test for pollutant gas levels has been introduced into the annual MOT test for roadworthiness.

The Chemistry of Ozone **down here** when NO_2 and HCs are present. During daylight O_3 concentrations are controlled by three chemical reactions:

1. $NO_2 + SL = NO + O$ (a free O atom is made)
2. $O + O_2 + m = O_3 + m$ (ozone is created)
3. $NO + O_3 = NO_2 + O_2$ (ozone is destroyed)

(SL is sunlight, m is a molecule of any other chemical)

All three reactions take place very quickly so that O_3 formation is balanced by its removal except when it is very sunny and Hydrocarbons are present (e.g. from idling engines in a traffic snarl-up). With HCs present NO can be oxidized (have O added) into NO_2 without O_3 being involved as it is in reaction 3 above. Thus the amount of O_3 increases as well as the NO_2.

Graph showing the pollution peaks during the course of one day in a large town.

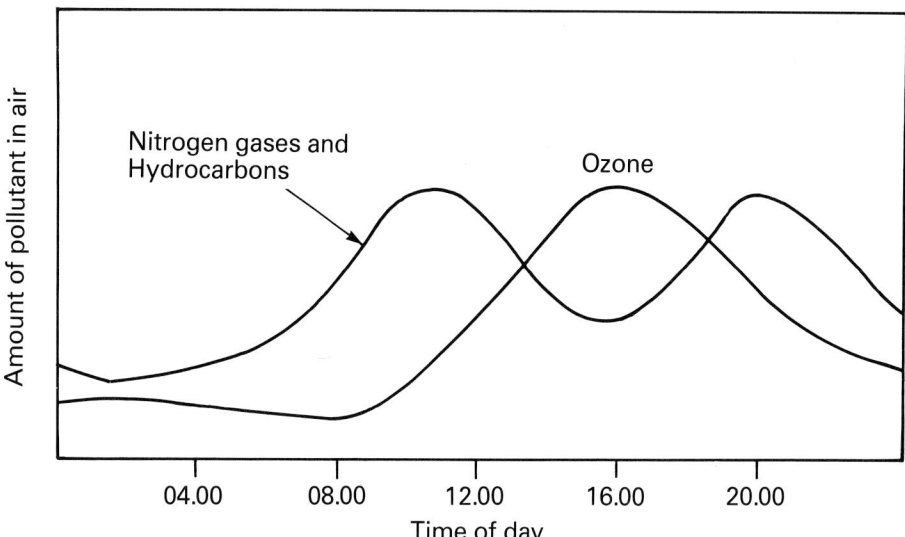

population lives. It is true that ozone is not one of the main greenhouse gases, CO_2 is the most important, but ozone is a more 'powerful' absorber of heat than CO_2 – it is 2,000 times 'better' at it. Since we are also concerned about CFC gases with ozone destruction in the upper atmosphere, it is worth noting that these gases are also greenhouse gases and they have an even greater effect than either CO_2 or ozone – between 10,000 and 25,000 times more than CO_2. Ridding the world of the pollutants which cause the ozone problem, would, at the same time, help to prevent the Greenhouse Effect.

One interesting phenomenon of the creation of ozone at ground level is that it peaks during the afternoon period, a delay of several hours after the pollutant production has peaked. (See the graph above.) Since O_3 is a secondary pollutant it will not be formed until the sun has had time to act upon the primary pollutant gases spewing from the vehicles during every urban rush hour, sometime between 8 a.m. and 11 a.m. It takes a few hours for the maximum amount of O_3 to be formed. Another seemingly odd fact is that places well away from the urban centres where the traffic volume is highest may have higher concentrations of ozone than the city centres. Quite simply, any wind will move the O_3 away from the centre out into the suburbs or the surrounding countryside. Ironically the centre

of the traffic congestion may not be the main site for maximum O_3, although it will be the worst for fumes from exhaust pipes!

Aircraft play their part in providing the pollutants which form ozone **down here**. The role they play is explained on page 55.

Air pollutants from various types of electricity-generating power stations

Fuel Used	Efficiency	Emissions		
		NO_x	SO_2	CO_2
	(per cent)	(grams per kilowatt hour)		
High-Sulphur Coal-Fired (without SO_2 removed)	36	4.3	21.1	889
High-Sulphur Coal-Fired (with SO_2 removed)	36	4.3	2.1	889
Low-Sulphur Coal-Fired (with improved furnaces)	32	0.3	1.2	975
Oil-Fired	33	1.4	1.6	794
Coal Gas	38	0.2	0.3	747
Natural Gas	43	0.3	0	416
Natural Gas with advanced equipment	55	0.03	0	331

(Source: 'State of the World 1992')

This chart shows how the different types of power station pollution vary with the fossil fuels used and the methods of generation. A similar chart for alternative power methods would show no pollution emissions.

MONTHLY MEAN

Chlorofluoro-carbons

Several types of Chlorofluoro-carbons (CFCs) exist, and each is involved in a particular indus-trial use. CFC 11 is well-suited to make foams: it can be blown into polyurethane and other materials where it becomes trapped as a bubble. Since CFC is a poor conductor of heat, it is a good insulator which prevents heat escaping. So it is used to make cups for hot drinks, insulation panels for ceilings and in freezers. Because it is rigid, the panels themselves help with the con-struction of buildings and equip-ment. Another advantage is that the bubbles can be made to vary in size so that car panels for the instrument fascia can be made firm on the outside but soft on the inside – a safety feature not to be underestimated for reducing road casualties. But break the foam and the gases escape! CFC 12 – like the other CFCs, non-poisonous, non-flammable, non-corrosive, stable and cheap to make – was used as a cooling agent in air-conditioning and refrigeration. But when old equipment is broken up, the CFCs escape! As an aerosol propellant, CFCs are ideal. They not only force out the

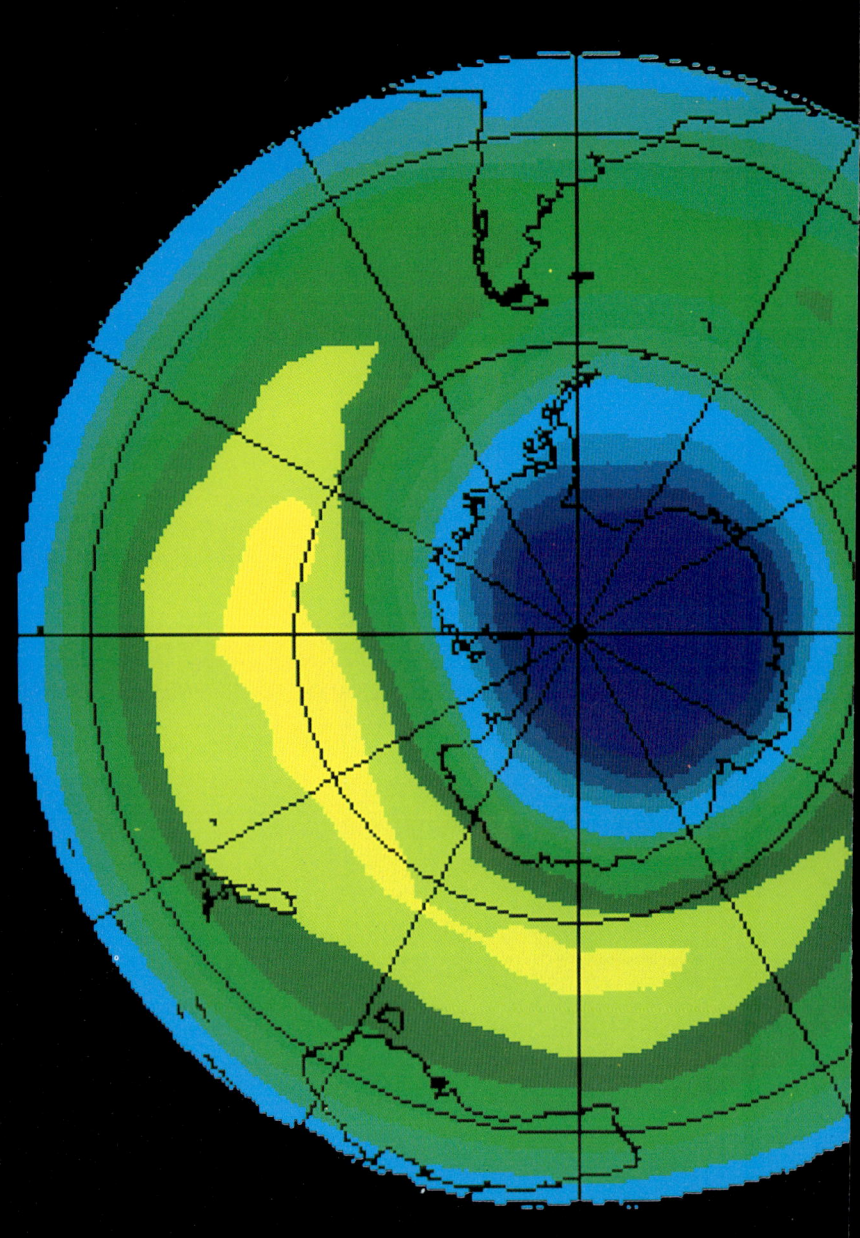

OCTOBER 19

TOTAL OZONE

contents but they can ensure that the aerosol droplets are the right size – too big and the droplets will not penetrate the lungs. However the CFC which pushed the droplets out of the aerosol escapes into the air, to add to the other CFCs going up there. In the electronics industry, CFCs can be used to clean oil, flux and grease from the circuit panels used in computers, televisions and so on. But unless the waste CFC is collected, it escapes into the air!

500 D
O
B
400 S
O
N
300 U
N
I
200 T
S

100

NIMBUS 7 TOMS
TOTAL OZONE
NASA GSFC

This satellite spectrometer image shows the ozone 'hole' in October 1985. The black and dark blue colours indicate the areas with least O_3.

5

The problems with CFCs

They are cheap to make, at about $2/£1.20 for a kilo of CFC 12. This makes them popular with manufacturers who may have to pay much more for alternatives. CFCs last a long time before they are broken down by natural forces. CFC 11 lasts about 80 years, CFC 12 up to 140 years and CFC 115 nearly 400 years. This means that none of the CFC first produced will have disappeared by natural means by the year 2000. The main problem now is to control and eliminate the further production of CFCs and to ensure that all of the CFC now in use is not allowed to escape into space.

Halons

Halons are gases which contain bromine rather than chlorine: they have the property of being able to extinguish fire. Thus they are to be found in portable fire extinguishers and fixed fire-fighting equipment. Unfortunately, bromine is between 30 and 50 times more ozone *unfriendly* than the chlorine in CFCs. It is now estimated that halons may be responsible for a third of the ozone depletion, although CFC gases are 100 times more likely to be emitted into the air because of their greater use. But unlike CFCs, halons are likely to be stored for long periods before they are used. Tests, leaks, training and false alarms mean that double the amount actually used against real fires is released into the air.

Dealing with halons

The International Maritime Organization (IMO) is the United Nations organization responsible for regulating the world's commercial shipping. Since ten per cent of halon use is connected with such shipping it is important that the IMO should agree to control the problem. In 1992 a complete ban on using halons in fire tests on ships came into force, and no new halon installations will now be allowed except in special circumstances.

When the nations of the world get together to follow up the Montreal protocol again, halon control will be high on the agenda.

Ozone depletion in the ozone layer

Depletion means that the amount of ozone is getting less. One of the latest estimates from NASA is that in 1989 in the northern hemisphere – where most of the world's population is to be found – there was a six per cent loss of ozone compared with the amount ten years earlier for the zone between 53° N and 64° N. For the zone between 40° N and 53° N the loss was 4.7 per cent. What these figures mean is that a considerable extra amount of UVB radiation is reaching earth and for every 1 per cent extra ozone depletion, there will be a 6 per cent increase in skin cancers. In their 1991 report, the British Stratospheric Review Group said that the ozone reduction in the northern mid-latitude area was now 8 per cent. But the real shock of ozone depletion was the discovery of an OZONE HOLE over the South Pole as shown on the previous page. There is concern that a similar hole is to be found at the North Pole.

The existence of a hole was first suspected in 1982 by the British Antarctic Survey, later to be confirmed by the scientists of

Some of the CFCs and their chemical composition

CFC 11 $CFCl_3$ trichlorofluoromethane
CFC 12 CF_2Cl_2 dichlorodifluoromethane
CFC 115 CF_3CF_2Cl chloropentafluoromethane
Halon 1211 Bromochlorodifluoroethane
Halon 1301 Bromotrifluoroethane
Halon 2402 Dibromotetrafluoroethane
The CFCs are made by various chemical firms under different names for example: FREON (Du Pont) ARCTON (ICI) ISCEON (RTZ Chemicals) FRIGEN (Hoechst) KALTRON (Kali Chemicals)

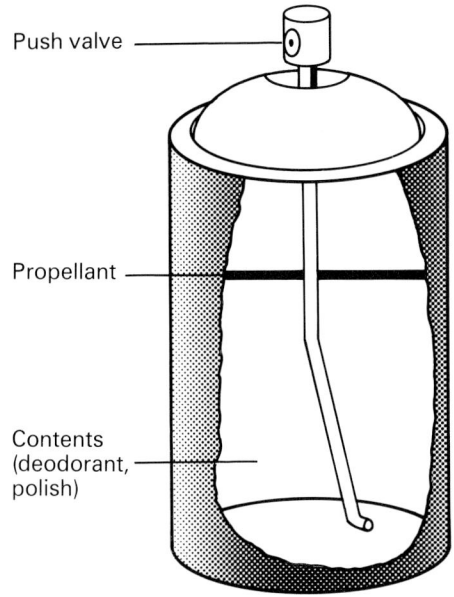

How an aerosol works.

Push valve

Propellant

Contents (deodorant, polish)

the American space programme (NASA) (see page 28). Certainly over 20 years ago, warnings of an 18 per cent depletion (thinning) of the ozone in the ozone layer were being made. In the early 1990s the hole previously reported in the ozone layer is still increasing in size, confirming the worst fears of scientists.

... and the cause? ... almost certainly CFC and halon gases.

Chlorofluorocarbons (CFCs) were created over 60 years ago, in 1928, by chemists working for the US General Motor Company. Their greatest value to industry was that they could be made to do things without having any other effect. CFCs are stable (slow to change) and inert (will not affect or harm anything they touch in factory processes). In the 1930s their use was limited to providing the cooling liquid for refrigerators rather than ammonia. It was not until the 1960s that they started to be used widely for other purposes. Halons, too, were used in industry. The difference between CFCs and halons is that chlorine makes the first and bromine the second.

In 1973, Dr Mario Molina and Professor Sherwood Rowland published the results of their research, the unpleasant and very worrying discovery that CFCs destroyed ozone. Because CFCs were so stable, they were not, like other chemicals, destroyed as they ascended into the sky through the atmosphere. Now it has been found that CFCs and halons are 'greenhouse gases' and are one of the causes of global warming.

Destruction of ozone by CFCs

When a UV ray hits a CFC molecule it causes a chlorine (Cl) atom to escape. This then acts upon the ozone molecule to form oxygen, and captures the free O atom to make chlorine monoxide (ClO). This ClO is attracted to another free O atom so that the two O atoms join up to make ordinary oxygen gas leaving the Cl atom by itself and able to attack the next ozone molecule. Once more the Cl breaks up the ozone: the process repeats itself, maybe 100,000 times before the free Cl atom is finally neutralized. A chain reaction has taken place. As a result, up to 100,000 molecules of ozone have been destroyed – 100,000 protectors of Planet Earth. Naturally-produced chlorine, together with other chemicals such as methane and nitrogen oxides, has always reacted with the ozone **up there**, but there was a natural balance of 'make and take', until man-made CFC came along and methane levels also increased.

A molecular diagram showing how O$_3$ is destroyed by CFCs.

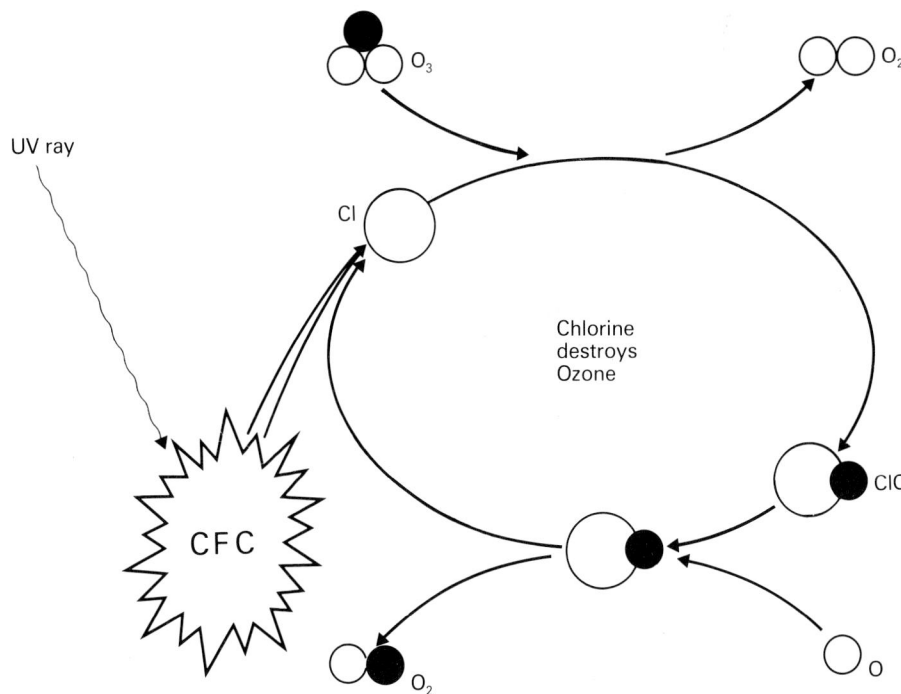

The hole in the ozone layer: a cautionary tale

The British Antarctic Survey (BAS) has been measuring ozone for about 30 years. Until the start of the 1970s the data discovered was published for all to see. As a result of economy measures by the British Government, the BAS did not have the money to issue the ozone statistics so they remained unexamined by researchers. It was not until about ten years later that Jonathon Shanklin and Joe Farman of the BAS started to look at the 'sleeping' ozone figures: they discovered that the level of ozone over Antarctica was getting lower. They were able to link what they knew about CFCs with this lowering of ozone.

In May 1985 they published an article in the journal *Nature* which suggested that man-made chemicals were destroying the ozone layer. American scientists wondered why their own satellite, which had gathered evidence on ozone, had not shown this. When they looked at the data again it did, in fact, prove the same point – ozone was being depleted in the ozone layer. From the figures they

had already received from the Nimbus satellite, they were able to draw maps which revealed the 'hole' above the Antarctic.

The question was, why had they not discovered this earlier? A simple, but serious, explanation was forthcoming: the computer processing the information had been programmed to ignore low ozone values because no-one believed that levels could ever fall so low. Now in the 1990s, the question is simple: if the world had known of ozone depletion ten years earlier, what steps could have been taken to avoid the disastrous consequences we now face? We cannot answer that question but we can conclude:

- politicians who control the nation's finances should not be tempted to cut back on the money given for basic research, however dull and useless it appears to be;
- scientists should remember not to jump to a conclusion before an investigation has begun.

What was discovered?

At first, many scientists dismissed the proven 'hole' as being due to the polar Vortex, and therefore not man-made. Later research by the British Antarctic Survey and by NASA suggested otherwise. In September 1987, following an important conference in Montreal about O_3 depletion, a NASA report stated three things:

1 In the Antarctic spring (September/October) a hole existed in the ozone layer that was as big as the USA and as high as Mount Everest.

2 In places, only two-and-a-half per cent of normal levels of O_3 was present.

3 In the higher altitudes, levels of chlorine monoxide (ClO), which comes from CFCs, were 1000 times higher than normal. Obviously, these could not have been formed by CFCs from uninhabited Antarctica, they must have 'blown' in from the north. Since that time, a hole in the O_3 above the Arctic has been identified, but it is much smaller than the one in the south.

The earth's atmosphere

The layer of gases which surrounds the earth and which eventually merges into outer space is known as the **atmosphere**. That part nearest to the earth's surface is known as the **tropo-**sphere, the outer part the **ionosphere** and in between the **stratosphere**. Three-quarters of the gas in the atmosphere is nitrogen, the remaining quarter is almost all oxygen with only about one per cent left for all the other gases, including carbon dioxide.

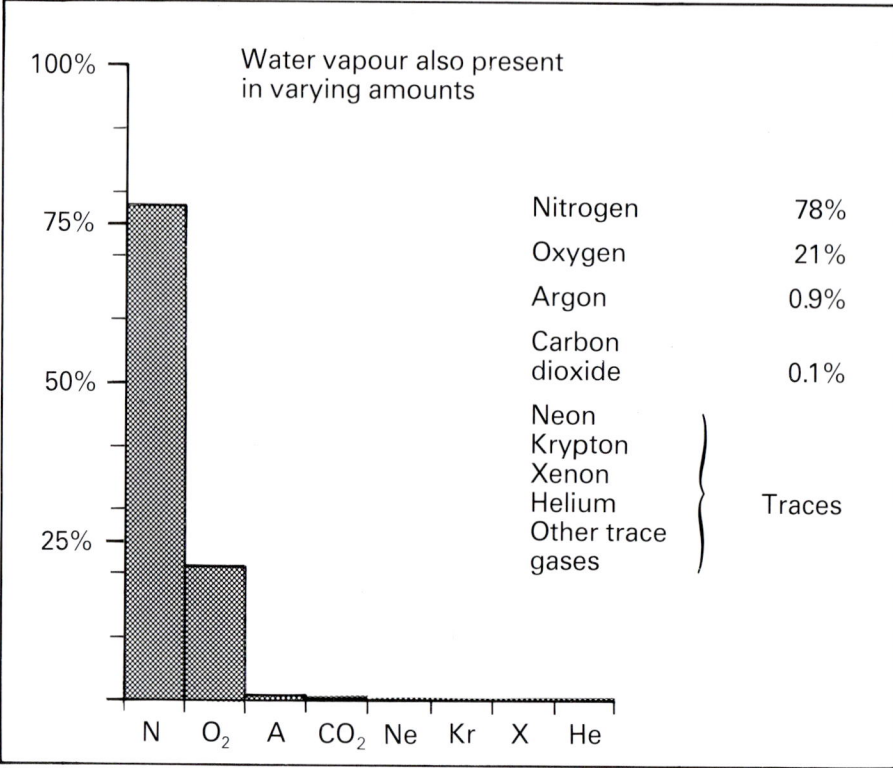

Gaseous content of the air we breathe.

Antarctic Vortex

The stratosphere above Antarctica is unlike any other part of the earth's atmosphere. From September to November it is dominated by the Polar Vortex, a tight whirlpool of winds, known to geographers as the 'Roaring Forties'. Thus the air above the South Pole is cut off from the surrounding air. This makes a space above the pole shaped like a tall pail. Remember there is no sunshine at the South Pole in winter (June, July, August) so what happens in the atmosphere there is unlike anywhere else on earth. Chlorine atoms are released from the pollutants carried there from their source elsewhere in the world. Because the Vortex winds have isolated the area, the ozone hole is mostly confined to Antarctica.

Australian walkabout

The Aborigines, the native population of Australia, are well known for the fact that on occasions they go 'walkabout' – they embark on long treks. Australian scientists at the Bureau of Meteorology in Melbourne have found that at the time of the break-up of the ozone hole above the South Pole (close to Australia in geographical terms) it had gone 'walkabout' to such an extent that in mid-December 1987, ozone levels above Melbourne dropped by 12 per cent. The conclusion was that the 'hole' had drifted slowly about 2,500 km (1,500 miles) from Antarctica at a speed of about 300 km (200 miles) a day.

Dr Joe Farman suggests this walkabout has caused the hole to spread over populated areas of Australia, New Zealand and South America with the accompanying danger to people from the extra UVB and possibly even UVC rays.

Radon

This gas is naturally produced from rocks and minerals which have a radioactive content such as granite or uranium. People living in houses built over such deposits or constructed from blocks of 'suitable' rock may inhale radon. Exposure to sufficient quantities over a long period of time can cause lung cancer. (See *Radiation and Nuclear Energy* in this series for a full explanation.)

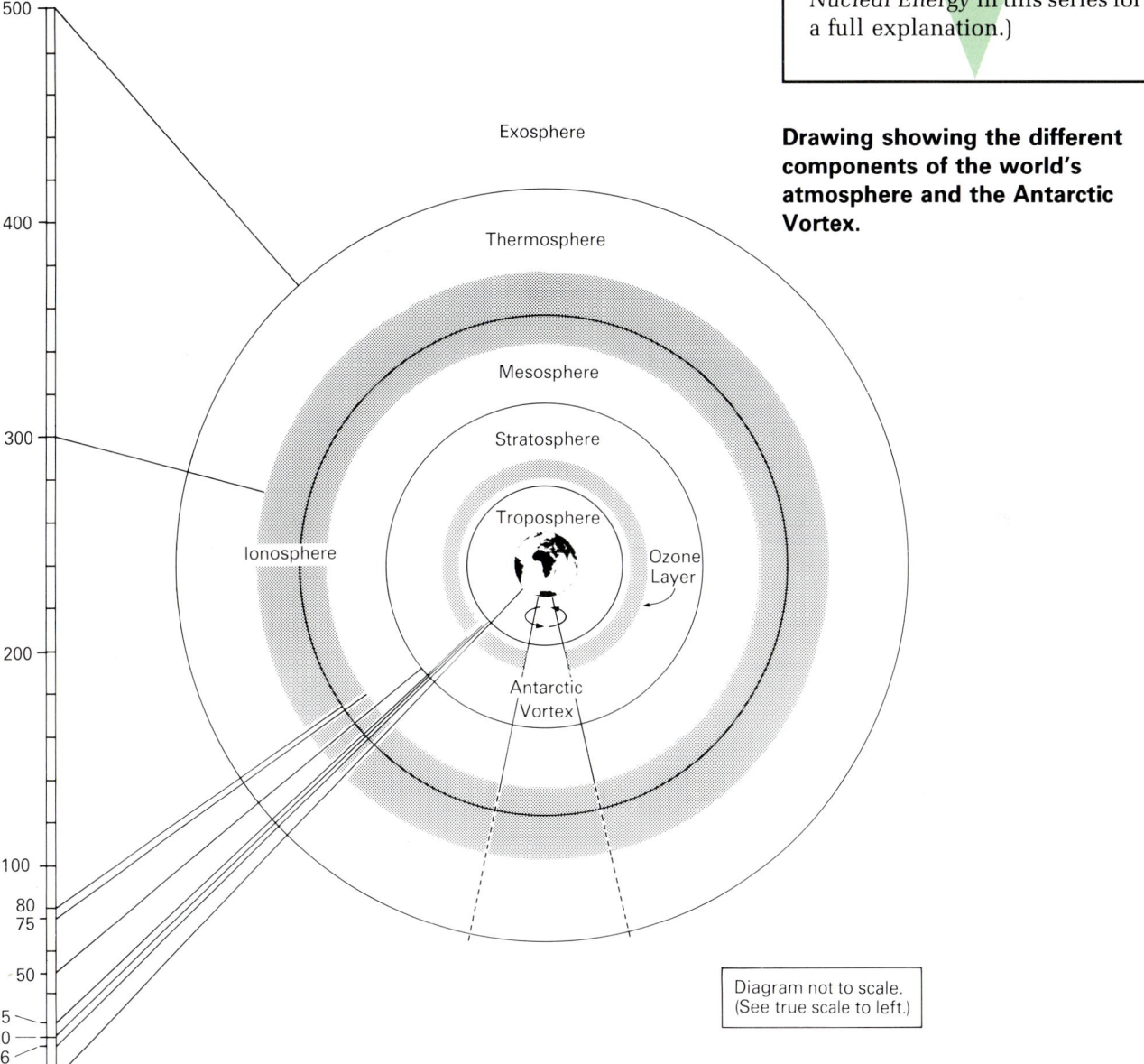

Drawing showing the different components of the world's atmosphere and the Antarctic Vortex.

True scale in kilometres

OTHER POLLUTION IN THE AIR

In the introduction I said that although this book was concentrating on air pollution which caused ozone problems it was not sensible, or even possible, to isolate one pollutant entirely. A headline in the *Guardian* of 18 May 1990, just after the momentous political changes in Eastern Europe read: 'Still, the darkness at noon.' It referred to the town of Copsa Mica in Romania, where the local carbon factory spewed out black dust and clouds of industrial smoke most of the time. It mattered little to the people of Copsa Mica whether the air was polluted with ozone, the whole cocktail of gases and soot choked their breathing and, like the inhabitants of Krakow in Poland, 'made their mood as bitter and their talk as acrid as the clouds of sulphurous dust that pour over Krakow from the Nowa Huta steelworks'. With 1,000 tonnes of carbon monoxide emitted *every day* from its 450 chimneys, together with the thousands of tonnes of sulphur dioxide and dust annually, it is little surprise that on average, men and women live six to eight years less than people in western Europe or the USA. In Krakow there is a four times greater chance of dying of cancer as in the UK or the USA even though, by 1992, the shutdown of the worst polluting coke-making units had reduced metallic dust, carbon monoxide and other gas emissions. Steel production had fallen by half, which had brought CO down by a third: the aim is to reduce dust and SO_2 to 3,000 tonnes, each, of exhaust annually.

Pollution in the West

But things can be as bad in the western world. In the USA during the summer of 1988, half of the country suffered a heatwave

The sun is barely visible through the pollution in the town of Copsa Mica, Romania.

which lasted almost two months. In New York the newspapers stopped mentioning days when the temperature rose above 32°C (90°F) – but what they did report was the unbearable foulness of the air. The *New York Times* wrote on 14 August 1988: 'The suns and seas and sins of man have combined to transfer New York life into a seemingly endless slog through simmering broth.' The broth was made up of vehicle exhaust, industrial fumes and the gases rising from the rotting rubbish of thousands of homes, offices and factories. Pollution was so bad, and the air along the bathing beaches of the Hudson river stank so horribly, that people avoided the area.

In a simple way let us look at the main pollutants that might be found in various forms of air pollution. If they were inhaled for a long enough time, what would the effect be on human health? The air we breathe is made up of the 'normal' gases such as oxygen and nitrogen, plus gases, liquid (mainly water) and solid particles some of which we classify as pollutants. In dry areas, for instance, the air will contain sand grains and in industrial areas unburnt particles or soot, will pollute our intake of breath. To try to simplify matters I have listed the various pollutants into categories, although technically it could be argued that some could come under another heading.

Gases

Many of the gases which cause air pollution have been mentioned already in connection with ozone. The less common gases are not listed here.

1 Sulphur dioxide SO_2

SO_2 is formed when the sulphur in coal, oil and other fossil fuels is burnt and mixes with the oxygen in the air. It is a colourless gas with a sharp smell. A great deal of SO_2 is produced naturally by volcanoes and other volcanic activity. It hurts eyes and lungs and is the major cause of acid rain when it has reacted to form sulphuric acid.

2 Hydrogen sulphide H_2S

This is the gas which smells of rotting eggs and provides the main ingredient for a 'stink bomb'. It is produced during oil-refining and other industrial processes. In nature it is formed when once living (organic) material rots, and also comes from sewage. It causes nausea and hurts the eyes.

3 Carbon monoxide CO

Another gas produced when fossil fuel is burnt. In particular it is made when petrol explodes in motor vehicles. From a health point of view, it is the most damaging gas and if enough is inhaled it causes death. It acts by reducing the amount of oxygen which blood can carry around the body. A shortage of oxygen damages the heart and, ultimately, stops it beating altogether. If people are already suffering from heart disease, breathing in CO quickly has an adverse effect.

The diagram indicates some of the sources of many pollutant gases.

4 Carbon dioxide CO_2

CO_2 is not an actual pollutant, hardly surprising when humans breathe it out of their lungs and plants take it in during photosynthesis. It is the important greenhouse gas which leads to global warming. It is formed by the incineration of fossil fuels and other material.

5 Nitrogen oxides NO_x

There are many combinations of nitrogen and oxygen which can be produced by burning fossil fuels, spreading nitrate fertilizers and from industrial processes. Much is naturally produced. Some results from photochemical actions. NO_x are dangerous to people with breathing problems such as asthma, bronchitis and emphysema.

6 Hydrogen chloride HCl

This gas is given off when waste material is burnt. It affects both the eyes and the lungs.

7 Silicon tetrafluoride SiF_4

Chemical factories are the main source of this pollutant gas. The main harm it does is to the lungs.

8 Chlorine Cl_2

Another gas which enters the exhaust fumes from a chemical factory. It reacts chemically with water in the atmosphere to form hydrochloric acid and so contributes to acid rain. If enough Cl_2 is inhaled it will damage the lining of the nose.

9 Fluorine F

This gas is an exhaust product of metalmaking factories. It can do considerable harm if inhaled in large doses.

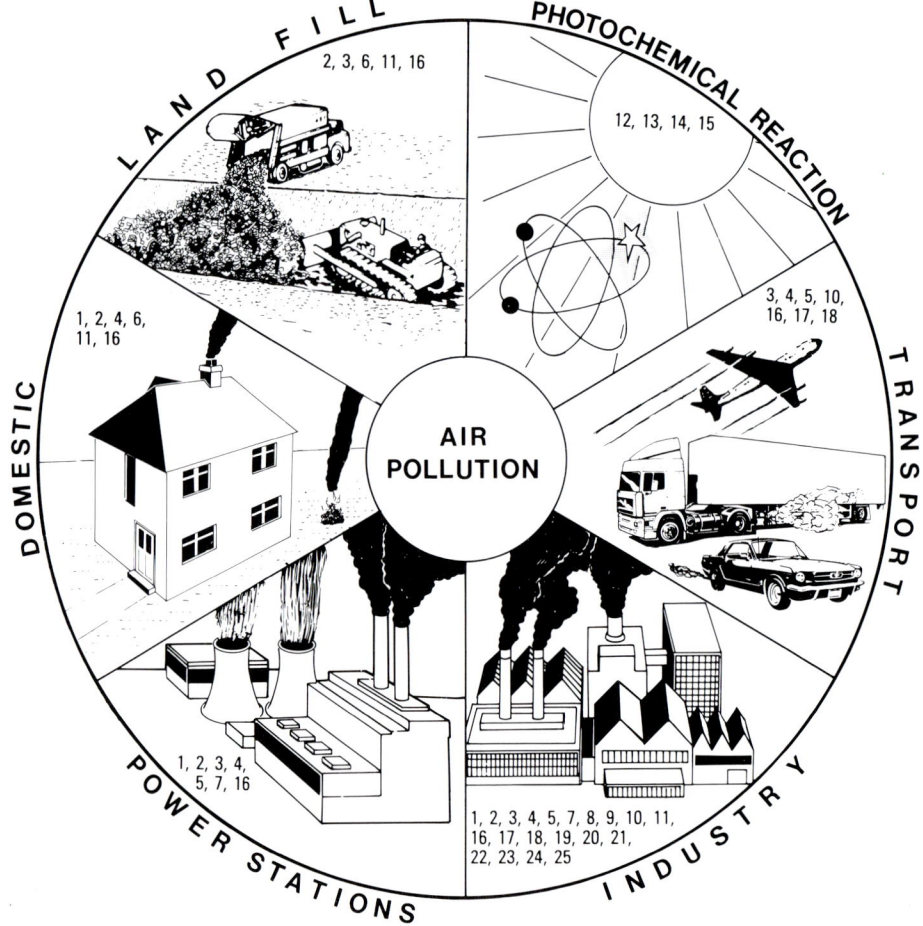

10 Formaldehyde HCHO
In liquid form this chemical is used to preserve biological items such as the internal organs of people. As a gas present in vehicle exhaust and industrial fumes it is damaging to eyes and lungs.

11 Chlorofluorocarbons CFCs
Manufactured for their stable and inert qualities, these gases cause global warming and ozone destruction as discussed in this book.

Photochemical products

The action of the sun on certain gases gives rise to pollutants. Of these, ozone is one of the most important, as explained elsewhere in this book.

12 Peroxyacetyl nitrate PAN
From the sun's action on NO_x and hydrocarbons. PAN and ozone are the two main ingredients of photochemical smog. PAN causes painful irritation of the eyes and affects lungs adversely.

13 Hydroxyl radical OH
This, too, forms from NO_x and HCs. Although OH is damaging to health its conversion from greenhouse gases means that global warming is reduced.

14 Sulphuric acid H_2SO_4
Photochemical reaction with SO_2 gives rise to sulphuric acid which is extremely damaging to lung tissue and an acid rain maker.

15 Nitric HNO_3 and Nitrous HONO acids
These are photochemical products from nitrogen dioxide. As well as contributing to acid rain, they harm lungs.

Volatile Organic Compounds VOC

(also known as Hydrocarbons)
There are many VOCs all of which are harmful to health. The best known are methane, benzine, ethanol, butane, propane and toluene.

16 Methane CH_4
This is formed from rotting vegetation, decaying rubbish, sewage, animal waste and is part of the flatulence, or wind, from all animals, especially cattle. Termites produce as much as five litres of methane a day from each termite mound. This gas is one of the greenhouse gases and as far as human health is concerned it affects peoples' breathing. Like other VOCs, it could be used as a fuel to provide power.

17 Benzine C_6H_6
This comes from vehicle exhaust and oil refineries. It causes leukaemia (cancer of the blood) in people. VOCs are generally carcinogens (can cause cancer).

Metals

Metals are emitted from industrial processes and, in the case of lead, from petrol motors using fuel with lead added. The main metals are:

18 Lead Pb
Stunts growth, damages brain cells and leads to high blood pressure.

19 Cadmium Cd
Damages lungs, kidneys and weakens bones.

20 Mercury Hg
Causes severe brain problems leading to body trembling, fits and death.

21 Magnesium Mg
Amongst other things it is thought to be one cause of the body shaking in people with Parkinson's disease.

22 Nickel Ni
One of its effects is to cause lung cancer.

23 Arsenic As
This can cause lung and skin cancer.

Solvents

24 Carbon tetrachloride
This is usually used in liquid form but it vaporizes easily to form a gas. It is used as a solvent – a cleaning agent often used in the dry cleaning of clothes. It is estimated that eight per cent of ozone depletion is due to this gas. It is to be phased out of use by the year 2000.

25 Methyl chloroform
This is another solvent used in industrial processes. It will also cease to be used by the year 2000. Five per cent of ozone layer depletion is attributed to methyl chloroform.

Particulates

This is the name given to all of the unburnt particles which are in the air, such as soot. They contribute to the formation of smog. Inhaling these particles damages the linings of the nose and lungs.

DOWN HERE – THE CURES

Wind turbine fields like this one in California, could offer a real alternative to fossil fuels.

If ozone pollution is caused by the reaction of sunlight on various pollutant gases, it follows that the only cure can be to stop either the sunlight or the pollutants. Since the first step is not possible, concentration of effort must be reserved for resolving the problem of emissions of harmful gases from industry, domestic sources and, most important of all, from the exhaust pipes of motor vehicles. That this can be done is not in question. Whether people have the will to do so is the hurdle that has to be overcome. It is not just a matter of inventing alternative non-polluting processes, or spending more money; people must be prepared to alter their whole way of life.

Incentives for change

It may be possible to force people to change their habits and to do what they can to prevent pollution. New laws are one solution, taxation is another weapon which can be used to persuade people into a different lifestyle. Making one method of doing something more expensive than another will inevitably lead to a change. Politicians must be encouraged to create taxes and pass laws which will help to combat pollution. To take a case in point, the addition of lead to petrol improves engine performance but it does mean that the metal is emitted into the air to be breathed in by passers-by. Lead is damaging to health, in particular to the brain development of children. In the UK, the government has reduced the tax paid on unleaded petrol by about five per cent compared with leaded petrol. This has resulted in many drivers switching to unleaded fuel even though for some it involved the cost of adjusting the car engine to take this type of petrol.

Direct precautions may be taken by wearing a protective mask on days when the ozone level is high. This is hardly a happy state of affairs and it also means the public will need to know the forecast for ozone levels.

There is no doubt that the only complete 'cure' is to prevent the production of pollutants in the first place, which is easier said than done! Pollution-free electric vehicles already exist and are especially useful in a journey where they have to keep starting and stopping and where the total mileage to be covered is not high. In Britain milk is delivered daily to homes and the empty bottles collected for re-use. Electric milk floats are in common use for this, powered by several large and heavy batteries. Battery weight is a problem as some of the energy used is to move the weight of the power unit. A more serious difficulty is the need to recharge the batteries. Although the vehicle has no exhaust fumes, the electricity used for recharging has to be generated at a power station. If it is a generator powered with fossil fuel, (gas, coal or oil) pollutant gases will be emitted – in other words the electric vehicle causes pollution indirectly, but still far less than if it were petrol driven.

Alternative energy sources

Electricity power stations do not have to be fossil fuel driven: they could be just as effective using alternative sources of energy such as wind, wave, solar or water (hydro). These are almost pollution-free, but such generators exist in the world in insufficient numbers to make much difference to the overall reduction of pollution. Nuclear power stations are pollutant free but there is always the danger of radiation leaks from plants which rely on FISSION whereby the atom is split to produce the heat needed to change water into steam and thus into electricity. FUSION is a method whereby the atoms are fused together and not split which is both pollutant and radiation free. This method imitates the way the sun works. As yet fusion is only at the experimental stage. In November 1991 the process was made to work for just two seconds – not long but enough to show that the basic idea is possible. The success was at the Joint European Torus (JET) laboratory in Oxfordshire, England. Scientists are hopeful that this clean nuclear power will be used for power stations early in the twenty-first century.

Since it is the fuel which causes the pollution, why not use something which is pollution free? Hydrogen is just such a fuel. When burnt, hydrogen produces no harmful exhaust, only water. It is also inexhaustible since it releases the same amount of water as is used to create the hydrogen in the first place. This is a completely renewable energy cycle. Mazda, the motor manufacturer, has already produced the HR-X city car which is powered by a rotary engine which runs on hydrogen. It also has electric batteries which are recharged by the engine after being used to accelerate the car. The car is still only in the trial stages but will be available for general use before the year 2000.

Catalytic converters

With so many vehicles already on the roads and large numbers of aircraft in the air, practical ways of reducing pollution must be introduced now, rather than

Buses in Sichuan province, China, use biogas (Methane) instead of diesel.

aiming for complete cures. Catalytic converters are honeycomb-filled bulges in the exhaust systems of a vehicle which destroy the pollutants and turn them into harmless gases, from an ozone production point of view. CO_2 is created which may not add to the ozone levels but does increase global warming. Lean burn engines create less pollution than engines which use a richer petrol to air mixture. Nitrogen oxides are at a minimum when the mixture is about 18 parts of air to one of petrol. Whereas 'cats' have to be replaced every three years or 50,000 miles of motoring, lean burn engines remain effective throughout the life of the car.

Even without these technical improvements there would be less pollution if engines were kept in good order, people drove at slower speeds or, better still, left the car at home and used public transport. When there are long tail-back jams on motorways, freeways or autobahns, traffic signals could be introduced, telling drivers to switch off their engines and wait for the road to clear. Engines that are just ticking over produce the most pollution so this precaution would cut down the gases emitted by a considerable amount. As we have seen, some cities ban vehicles from the city centres when pollution levels are too high.

One way to persuade drivers to travel fewer miles is to make fuel so expensive they cannot afford to drive far. Some economists estimate that $10 or £5 a gallon is about the price to deter motorists from all but essential car use – although people with lots of money would be able to beat the system which would defeat the object. Where industry is concerned, the above arguments for electricity still apply. Energy conservation is the main way of preventing pollution. Saving on wasted energy by proper insulation and efficient machinery would lead to the use of less energy and therefore the creation of less pollutant gas.

Even if it were possible to halt the production and the release of ozone-damaging gases the problem of those already in the ozone layer or on their way up would still remain. Even if such an immediate stop takes place, these former releases will continue to cause damage, due to the chain reaction explained earlier (page 37). All of this does not mean that action taken at ground level is a waste of time. Before long a way to replace the lost ozone may be found which has a reasonable chance of success. In order to improve the pollution situation there will need to be a change in both manufacturing and commercial practice. In other words, there will need to be replacements for the damaging CFCs. The cooperation of those whose job it is to collect waste is needed to ensure that the unused or unwanted gases are destroyed in an environmentally safe way. Pressure for this to be done may also need to be supported by new rules and laws.

It is no good curing the problem in one part of the world, if the whole sequence of events is repeated somewhere else. The Chinese government, with its one thousand million citizens, wishes to provide every home with a refrigerator. It will be useless for western countries to

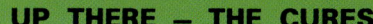

alter their refrigeration method to eliminate CFCs if China introduce new refrigerators using the old process. Help, especially money, will have to be given to overcome this problem. Money will also be needed to provide R & D (research and development) where alternatives to CFCs are concerned and where investigation into the replacement of lost ozone **up there** is contemplated.

It is possible to make some attempt to counteract the worst of the effects of the dangerous UVB rays reaching the earth as the result of ozone depletion. The Sun Smart campaign I described earlier is one of them. Other pollutant gases 'assist' the CFCs to damage the ozone layer – their presence could be depleted with the methods described in the previous chapter.

It is essential that the CFC gases in discarded refrigerators are recycled and not released into the atmosphere. Old refrigerators are piled up here, awaiting attention.

The Montreal Protocol

Montreal, in Canada, was chosen for the 1987 world conference on the problem of the depletion of the ozone layer. As a result the Montreal Protocol was published, which set a target of reducing the use of CFCs by 50 per cent by the year 1999. In March 1989, at a London conference on ozone depletion **up there**, it was agreed that a 90 per cent ban was needed if any significant difference was to be made to the amount of ozone being lost. The only major objecting country was the then USSR. In the early 1990s, that country underwent a total political change and it remains to be seen what effects this will have on CFC control. The Helsinki (Finland) Declaration on the Protection of the Ozone Layer in May 1989, agreed by nearly 100 countries, set the year 2000 as the date by which all CFCs should have been phased out and in 1990, other ozone-destroying chemicals were added. In February 1992, the USA brought forward their CFC ban date to 1995. In Britain the company which made ten per cent of the world's CFCs, ICI (Imperial Chemical Industries), also set a target date of 1995, to cease making CFCs. Hopefully others will follow these leads. India, China and other developing countries suggested that the richer western nations should finance 'Phase Out Aid' to pay for the extra cost of new technology to reduce their use of CFCs. In June 1990 the 'Montreal Protocol' nations agreed to set up a $250 million fund (about £140 million) for this purpose. More international co-operation will be needed, for the problem of ozone depletion is not a country by country issue – it is a global problem.

Alternatives to CFCs

Agreeing to phase out CFCs is one thing – how to do it is another. It is possible to find simpler ways of using items 'powered' by CFCs: for example anti-perspirant aerosol sprays could be replaced by roll-on tubes, and aerosol shaving cream dispensers by pump action cans. But a return to simpler living is not acceptable to some people. CFCs made life easier and more pleasant. Their contribution to efficient and fairly cheap refrigeration, air-conditioning, packaging, furniture and insulation is not something to be rejected lightly. The depletion of the ozone layer is a problem which, unlike the London or Los Angeles smog, is not directly obvious: smog can be seen, causes coughing, and is thoroughly unpleasant. Ozone layer depletion is somewhere else, cannot be seen, touched or smelt and its effects are not immediate.

There are some alternative chemicals already available. Butane can be used to propel aerosol sprays and the gas does not deplete ozone, but butane is highly flammable and is used as a bottled gas for fires, cookers and mobile flame torches. It is obvious that a butane spray could be very dangerous. Other alternatives have similar problems – nitrous oxide (laughing gas) is used in a cream dispenser: the label proudly boasts the propellant is not a CFC gas – but N_2O is thought to be a greenhouse gas though some recent research has questioned this. The way forward may be to find substitutes for CFCs which do not contain chlorine. This may be done by using hydrogen rather than chlorine so that instead of **chloro**fluorocarbons (CFCs) we will have **hydro**fluoroalkanes (HFAs). The HFAs are chemicals where the chlorine has been eliminated altogether or hydrogen has been substituted for the chlorine. Not only does chlorine, the enemy of ozone, disappear, but hydrogen makes the HFA unstable so that it reacts with water and oxygen to become harmless – remember it is the fact that the CFC is so stable that it manages to reach the upper atmosphere without being naturally destroyed. There are difficulties. It is more expensive to make HFAs as the chemical processes involved are more complex so the technology is more advanced. HFAs are not able to

A clever logo was chosen for the 1989 London conference on the ozone layer. The world is surrounded by the O of the chemical symbol for ozone with the $_3$ completing it.

do the same range of jobs as a CFC gas — in other words different HFAs have to be found for jobs previously done by just one CFC. Another difficulty is that special lubricants are needed for HFAs — previously CFCs worked well enough with simple mineral oils — more research means more expense.

Inevitably CFCs will still be used in manufacturing and other processes for many years to come, particularly in the less advanced countries. Then there is the CFC already in use which will be inside equipment when it is replaced. CFC waste will need to be collected in an organized way and arrangements made for it to be destroyed to prevent the gas from escaping into the atmosphere. In electronics factories, where CFC is used as a cleaning agent, even stricter control of the waste will be required than at present. The CFC will have to be recycled so that the loss into the atmosphere is minimal.

In some cases it is quite easy to substitute completely different products for purposes which involved the use of CFCs in manufacture. To take two simple but common examples. Egg boxes need to be light but strong — foamed polyurethane was ideal but it involved CFC in the foam- ing process — a return to the use of papier mâché material resolved this and also provided a use for recycled paper. Similarly, drinking cups made of waxed board or plastic can replace the CFC-made material. Many other examples could be cited.

As one of the largest users of takeaway packaging, MacDonalds now uses alternatives to CFC foamed materials in its boxes and cups.

GOOD NEWS – BAD NEWS

Place Sydney, Australia
Date August 1991 at the start of the Australian summer
Event Medical conference
Speaker Bill McCarthy, Professor and leading expert on skin cancer
Theme of speech Unless ozone repair measures are adopted, ultraviolet C radiation which is still blocked by the ozone layer, will quickly cause cancer to unprotected skin.

Spacesuits will be the sunscreens of the 21st century, when lethal doses of ultraviolet will begin to strike the earth, unless science can slow down the rate of ozone layer depletion. Just a few minutes of sunlight would cause you to burn. The ozone is the only protection we have from sunlight, which is lethal. If we extrapolate [draw out a conclusion] from what is currently happening with the ozone layer, if we can't prevent the deterioration, then ultraviolet C will hit the earth.

Are there no 'rays of hope' – not real physical rays but some cheerful items amongst the gloom? First and foremost, sunshine is good both for your body and for your mood. Second, there are always some scientists who take an opposing view, whatever the case. This is true of ozone layer depletion. I have collected together a few items which need to be considered before firm conclusions are drawn and any single thing blamed. It is these sort of matters which sometimes give politicians the excuse not to take action to prevent pollution damage.

Ozone destruction by aircraft exhaust and spacecraft

Nitrogen oxides, another source of ozone destruction, are one of the main emissions from engine exhaust including aircraft. It is doubtful if the exhaust of ordinary aircraft affect the higher regions of the atmosphere, although they contribute quite considerably to low-level air pollution. A jumbo jet is estimated to use in *the first five minutes of flight* the same amount of fuel, with the equivalent exhaust, as a single car will use in six years of average use! With supersonic aircraft it is different. The British/ French Concorde is the only commercial jet flying in the stratosphere. The military aircraft of the USA and the former USSR fly at even greater heights. All deposit their emissions directly into the ozone layer.

But what of the space shuttle 'aircraft'? One single flight of the US shuttle emits 187 tonnes of chlorine and seven tonnes of nitrogen chemicals before it reaches a height of 50 km above the earth. Another 840 tonnes of other gases are produced in a single flight: they may not be so destructive as ClO and NO_x but they are estimated to destroy their own weight in ozone. *One single shuttle flight can destroy 10 million tonnes of ozone . . . 300 shuttle flights could destroy the ozone layer altogether.*

Not everyone holds the same opinions on the relative dangers of ozone depletion. Here are some alternative views:

The ozone hole has always been there

Some scientists believe the hole has occurred regularly in times past but it has just not been discovered, and when it was found the fuss made about it by the media was out of all proportion to its real importance.

Global aircraft emission 1988

A WWF report by Earth Resources Research (ERR) stated that aero-engines flying globally emit the same amount of carbon as the whole of the UK, or nearly three per cent of the global carbon monoxide, nitrogen dioxide and sulphur dioxide as shown below.

Pollution	Commercial Aircraft	Military Aircraft	Total
Carbon monoxide CO	52	10	62
Hydrocarbons HC	154	29	183
Nitrogen dioxide NO_2	2,846	542	3,388
Sulphur dioxide SO_2	887	169	1,056
Water H_2O	184,831	35,206	220,037
Carbon C	133,522	25,433	158,955

(measured in 000 tonnes)

(Source ERR)

Destruction of the ozone layer – perhaps it is not all bad news!

It may well be that the ozone layer not only acts as a filter for the UV rays but it also prevents much of the radiated heat escaping back into space. In this latter role it keeps heat in the earth's atmosphere, therefore adding to the Greenhouse Effect which leads to global warming (see page 32). In December 1991 UNEP and the WMO reported that the ozone layer was so thin and badly damaged by CFC gases that its ability to prevent heat returning to space was severely lessened, which would lead to global cooling and not global warming!

Laughing gas – does it have the last laugh?

Nitrous oxide, N_2O, is commonly known as laughing gas because inhaling it is supposed to make people laugh. It is a gas used to anaesthetize patients, especially in dentistry. It has been accused of being a greenhouse gas, retaining warmth 160 times more effectively than CO_2. A new report from the IPCC reveals that a simple mistake in the original arithmetic has led scientists to a wrong conclusion.

Is the risk of skin cancer growing in the USA?

Sunburn test meters have been located all over the USA since 1979. They measured ultraviolet rays counting the amount of radiation filtering through the atmosphere. The result? During this period, the radiation amount actually DECREASED! It has been proved that the ozone layer has been depleted throughout this time so is there some scientific mechanism missing? Perhaps we don't have the story straight after all.

Nature at work – an alternative source of chlorine

A scientist at the Agricultural and Food Science Centre in Belfast, Northern Ireland claims a quarter of the chlorine in the upper atmosphere comes from fungi which causes wood to rot. Dr David Harper estimates that five and-a-half million tonnes of the chemical are produced annually when chloromethane is given off mainly by the fungi. Perhaps CFCs are being blamed for much that is a natural phenomenum. Chloromethane also comes from seaweed and from burning wood, making more natural chlorine atoms available to destroy the ozone layer. That is not all that nature contributes to ozone destruction **up there** – lightning flashes cause nitrogen and oxygen to combine thus creating more ozone 'enemies' to attack the ozone layer. Bacteria in the soil create NO_x and even the hydrogen and oxygen from water vapour increase the chemical constituents available to enable photochemical destruction of ozone to occur as described earlier on in this book.

Other factors

Some changes in the ozone layer are quite natural:

- areas of high atmospheric pressure are thought to produce 'transient' holes in the ozone layer – i.e. they come and go
- ozone levels are lower in winter due to lack of sunshine
- ozone levels are lower at night because there is no sunshine
- solar flares result in extra UV radiation which will affect the amount of destruction of ozone: other solar phenomena may alter the amount of UV radiation – sun spots or solar winds for example.
- volcanic eruptions may hurl a mixture of gases, aerosols (small liquid droplets) and dust high into the atmosphere depending on the angle of the ejecta (all the material thrown out).

El Chichón's ejecta (Mexico, 1982) went straight up, whereas that from Mt St Helens (USA, 1980) was low and sideways. Mt Pinutabo (Philippines, 1991) also had a significant impact high in the atmosphere, so much so by January 1992 the European Ozone Research Coordinating Unit and NASA, were giving out warnings of excessive depletion of the ozone layer over Northern Europe and America probably due to the extra effect of this latest eruption. In particular the chemicals ejected seemed to have greatly affected the amount of NO_x gases in the ozone layer with a consequent hazard to ozone levels. In 1982/83 similar low levels of ozone protection in northern latitudes were attributed to El Chichón, although the assistance of high-flying balloons and aircraft was not then available to provide detailed information.

The eruption of volcanoes such as Mt Pinutabo in the Philippines in June 1991, can seriously affect the depletion of ozone levels.

WHAT CAN I DO?

Signing the 1992 pledge to help the environment.

'You can pledge to take one or more of the following actions'

So began the publicity pamphlet raising support for the **Tree of Life – Pledging for the Planet Project** which aimed to take the voice of the ordinary people of the world to the Earth Summit in Rio de Janeiro held in June 1992. Each person was asked to make one or more pledges to act to 'Save the Earth' for, in the words of the organizers: 'if enough people make enough small contributions, lasting change becomes possible.' Millions of people responded, from all over the world. WERE YOU ONE OF THEM? If you were not too young to understand the project then there is no reason why you could not have participated. Hopefully you decided to keep one of the pledges listed by the project. They were:

1 I will use ten per cent less gas and/or electricity at home during the next year.
2 I will write to my bank manager urging the banks to write off debt owed by the poorer countries.

World leaders gathered for the Earth Summit in Rio.

3 I will cut my car mileage (or help cut my family's mileage) by ten per cent during the next year. I will walk, cycle or use public transport wherever possible.

4 I will recycle as much waste as possible at home and/or I will help to organize recycling at work or at school.

5 I will not buy any products made from tropical hardwoods.

6 I will try to buy products that are made without exploiting the people who produce them, or damaging the environment.

7 I will increase the amount of time or money I give to environmental organizations.

8 I will personally raise 20/50/100 pledge leaves for the Tree of Life in Brazil before 1 June 1992.

We have been looking at the problems of ozone – certainly pledges 1, 3 and 4 would directly help to reduce the pollution which causes the creation of ozone and some of the destruction of the ozone layer.

I can almost hear you say . . . 'But it's too late now – the Earth Summit is over, the Tree of Life is finished – I can't make a pledge' . . . and I can say in return, 'What's to stop you carrying out some of the pledges even now – we still have the problems to tackle – it's never too late to begin'.

Positive action

Make sure that any aerosols you use are not propelled with a CFC gas – nowadays they are usually labelled 'ozone' or 'environment friendly' on the can. If your family is changing the refrigerator make sure that the original one is collected so that the CFC in the unit is removed correctly. Find out whether packaging and disposable drinking cups are foamed with CFCs before using them. If the shop or snack bar doesn't know, write a letter of enquiry to the owner.

Use less electricity – the result will be less power station pollution unless yours comes from a non-polluting alternative energy station. The best way to avoid using so much is to ensure that your home, workplace, school or any other building you use regularly is properly insulated. Practising **energy conservation** is the best action anyone can take. Another personal action is to

In 1992 people were asked to make a pledge to help the environment and to send them in the form of a leaf for the Tree of Life at the Earth Summit in Rio de Janeiro.

cause less individual pollution: when travelling use public transport rather than a car or motorcycle and if you can, walk or bicycle instead. This is not always possible so it is important to use efficient cars fitted with catalytic converters or lean burn engines – **gas guzzlers are out**.

You can support organizations which are trying to convince governments to be environmentally friendly. Does your local council or community office provide a recycling service as is done in many towns and villages around the world? Write to your local representative to persuade her/him to support environmental efforts positively.

The Montreal Protocol, the London Conference on Ozone, the Earth Summit are all examples of international action to combat our worries about ozone. Have you noticed that newspapers and television weather forecasts now include pollution reports and probabilities? All this does show that people are becoming more aware of environmental problems: it is up to all of us – YOU included – to do our bit to keep the momentum going.

Q Can the loss of ozone **up there** be made up with the unwanted ozone **down here**?
A No. Ozone is a short-lived gas which breaks up easily. While some upper ozone moves down, ozone created near the ground never moves **up there**!
Q Will the ozone layer recover by itself?
A Hopefully in time it will – provided no more pollutants reach it.

As the organizers of the Tree of Life said: 'If enough people make enough small contributions, lasting change becomes possible.' Make your SMALL contribution a BIG one!

Countermeasure

Countermeasure is the term given to a process which deals with a problem after it has occurred. The use of catalytic converters in vehicle exhaust systems is one example, the Sun Smart campaign described earlier is another. Some people have suggested it might be possible to put the ozone that has been depleted back into the stratosphere or even to attack and destroy the damaging gases on their way up. At the present time this is more science fiction than science fact.

Some of the suggestions made have included:

- Using laser beams to destroy the CFC molecules as they reach the ozone layer.
- Having ozone-producing space factories pumping out ozone.
- Hurling capsules of ozone into space to explode and spill their contents.
- Having all high-flying aircraft fitted with ozone production equipment so that they can push extra ozone into the atmosphere.
- Sending small ozone ionizers into the atmosphere using helium filled balloons.

But all of this is very expensive and not easy to put into operation. Take the balloon idea for example. In March 1989 a newspaper story described a revolutionary system

Drivers can reduce exhaust gases by:

- Driving more slowly.
- Using fifth gear when possible.
- Keeping the engine well tuned.
- Buying cars which travel more miles per gallon.
- Walking or cycling for short journeys.
- Sharing rides to work.

to keep small ozone generators in space by automatically dumping ballast as the helium in the balloon carrier inevitably slowly leaked away. Nearly three years later the companies involved reported that 'the work we have carried out regarding floating solar-powered ozone generators into the stratosphere is still very much in the design stage . . .' emphasising the difficulties and the need for long-term research and development. The companies involved will be looking for millions of pounds from governments and national businesses to fund the scheme. The provision of this finance is a political decision – just how much money should be spent? Are we prepared to pay more tax to provide the money needed?

This page is headed 'What can I do about it?' Perhaps I should have said instead, 'What are you prepared to do about it?' At least you have been interested enough to find about OZONE – I hope it helps you to make up your mind.

GLOSSARY

Acid Rain
Rain that is a weak acid with a pH value below 5.0. The term is used for acid mist, acid hail, acid snow and even for 'dry' acid deposition.

Aerosol
Small drops of liquid which float in the air.

Air-conditioning
A system for cooling the air in closed areas.

Air Pressure
The weight of air pushing down on the earth. In weather forecasts it is referred to as 'low' or 'high'. It is measured in millibars.

Algae
Very small plants which often float on water, giving it a green colour.

Atmosphere
The air which surrounds the earth, made up of gases and water vapour.

Charcoal
Wood which has been 'cooked' and can be used as a substitute for coal.

Climate
The average weather conditions for a particular place.

Climate Model
A computer prediction of future climate.

Cloud
Droplets of water floating in the atmosphere.

Condensation
Water changing from a gas (vapour) to a liquid when it becomes colder.

Deoxyribonucleic Acid
DNA has been called the building block of life. The molecules which make up DNA carry the genetic code for cells – abnormal cells lead to abnormal results which cause odd things to happen. These changes are called mutations. Extra UVB damages DNA, thereby causing mutations in creatures and plants.

Dust Bowl
An area of wind erosion where the topsoil is blown away. The Mid-west area of the USA became known as the Dust Bowl in the 1930s.

Evaporation
Water changing from liquid to gas (vapour) when it becomes hotter.

Fossil Fuel
Something which has been formed millions of years ago, brought out of the ground and burnt. Once used, it has gone for ever. Coal, oil, natural gas, peat and brown coal are all fossil fuels.

Global Warming
Many scientists believe that the world will be warmer by between 1.5 and 4.5°C by the year 2050. This will be caused by the Greenhouse Effect whereby certain gases, mainly CO_2, retain some of the heat which should escape into space in a similar way to a greenhouse heating up when the sun shines. (See *The Greenhouse Effect* in this series.)

Insulation
Prevention of heat loss by putting in a barrier such as glass fibre 'wool'.

Montreal Protocol
An international agreement signed in 1987 to limit the use of CFC gases.

Mucous Membrane
A soft 'jelly-like' layer covering the inside of the throat and nose, and the outside of the eye.

Nuclear Fission
Splitting the atom to create heat.

Nuclear Fusion
Joining atoms together to create heat.

Nuclear Power
Using the heat from a nuclear reaction to create electricity.

Nutrient
That part of an animal or plant's intake which feeds its growth.

Photosynthesis
The process by which a green leaf makes starch and sugars for food. It results from the action of sunlight on the plant. CO_2 is absorbed during the process.

Phytoplankton
Minute algae (green plants) floating on top of the sea. Zooplankton (small creatures) 'graze' on it.

Radioactivity
The process of production of harmful radiation when atoms are split.

Radon
Gas given off when the radioactive element radium disintegrates.

Salination
The deposition of salt on the land after flooding.

Traffic Calming
Various measures taken to slow or lessen vehicle traffic such as 'speed bumps' or impeding movement with small traffic islands.

UN
The United Nations. Its headquarters are in New York.

RESOURCES AND ADDRESSES

Books

Atmosphere Oliver Allen, 'Planet Earth' Series, Time-Life Books, 1983

The Cousteau Almanac Jacques-Yves Cousteau, Columbus Books, 1981

A Dictionary of the Environment Steve Elsworth, Paladin, 1990

Dictionary of Environment and Development Andy Crump, Earthscan Publications, 1991

The End of Nature Bill McKibben, Penguin, 1990

Ozone and the Greenhouse Effect Stephen Trilling, Field Studies Council, 1990

Books in the 'Conservation 2000' Series, Batsford

The Acid Rain Effect Philip Neal, 1992

The Greenhouse Effect Philip Neal, 1992

Radiation and Nuclear Energy Joy Palmer, 1992

Books in the 'Considering Conservation' Series, Dryad Press

Acid Raid Philip Neal, 2nd Edition, 1988

Disappearing Rainforest Robert Prosser, 1987

The Encroaching Desert Norman Farmer, 1990

Energy, Power Sources and Electricity Philip Neal, 1989

Farming and Food Supply Derrick Golland, 1988

War on Waste Joy Palmer, 1988

The World's Water Joy Palmer, 1987

'Planning for Survival', 'Acid Rain', and other packs, International Centre for Conservation Education

Reports/Articles

'El Niño' *National Geographic* Vol. 165 No. 2, February 1984

'Into the Void?' A report on CFCs and the Ozone Layer, Kathy Johnston, Friends of the Earth, 1987

'State of the World 1992' Worldwatch Institute Report

Useful addresses

Department of the Environment
43 Marsham Street
London SW1P 3PY

Electricity Association
30 Millbank Place
London SW1

Friends of the Earth
26–28 Underwood Street
London N1 7JQ

Greenpeace
30–31 Islington Green
London N1 8XE

International Centre for Conservation Education
Greenfield House
Guiting Power
Cheltenham
Glous GL54 5T2

National Coal Board
Hobart House
Grosvenor Place
London SW1X 7AE

Warren Spring Laboratory
Department of Trade & Industry
Gunnels Wood Road
Stevenage
Hertfordshire SG1 2BX

USA and Worldwide

Anti-Cancer Council of Victoria
Keogh House
1 Rathdowne Street
Carlton South
Australia 3053

Environment Canada
4905 Dufferin Street
Downsview
Ontario
Canada M3H 5T4

Environmental Defence Fund
1616 P Street NW/150
Washington DC 20036

Friends of the Earth
218 D Street SE
Washington DC 20003

Renewable Fuels Association
201 Massachusetts Avenue NW/C
Washington DC 20002

UN Environment Programme
PO Box 30552
Nairobi
Kenya

US Department of Energy
Energy Information
Administration
Forretal Building
Washington DC 20585

US Environmental Protection Agency
401 M Street, SW
Washington DC 20460

World Resources Institute
1709 New York Avenue, NW/700
Washington DC 20006

Worldwatch Institute
1776 Massachusetts Avenue NW
Washington DC 20036

INDEX